Luzac's

Semitic Text and Translation Series,

Vol. XIX.

AMS PRESS
NEW YORK

THE
BANDLET OF RIGHTEOUSNESS
AN ETHIOPIAN BOOK OF LIFE
ALSO KNOWN AS
THE ETHIOPIAN BOOK OF THE DEAD

Edited with an English translation
By
Sir Wallis Budge. KT.
MA, LITT, DLITT; F.S.A., STAR OF ETHIOPIA,
3RD CLASS

With Sixty-Seven Plates

Re-published By
Research Associates School Times Publication
And Frontline Books
Co-published by Miguel Lorne Publishers
Distributed by Frontline Distribution Int'l Inc.
751 East 75th Street
Chicago, IL 60619

First published in London by:
Luzac & Co.
46 Great Russell Street, WC

Frontline Books Edition

ISBN 0-94839-087-5

BARBADOS • CHICAGO • JAMAICA • REPUBLIC OF TRINIDAD
AND TOBAGO

PREFACE

'ÊZÂNÂ, king of all Ethiopia, abjured paganism about A.D. 350, and, under the influence of political and commercial necessities, proclaimed Christianity the national religion of his Empire. He abandoned the use of the crescent and star on his memorial stelae, and the CROSS OF CHRIST took their place. But the greater number of his people were pagans, and they clung to their magical cults with characteristic tenacity. As Christianity made its way southwards from AKSÛM in the succeeding centuries, the people of non-Jewish origin became partially converted, but in spite of their outward professions and their acceptance of the doctrines of the Church of ALEXANDRIA and its rituals, they never wholly abandoned paganism. They did not, and could not, understand the higher spiritual truths of the Christian Religion, and the magician flourished side by side with the Christian priest. And the people generally preferred the former to the latter, for the former commanded the celestial powers to do his bidding by means of his spells and names and words of power, whilst the latter could only entreat them to help him through petitions and prayers. The Ethiopian craved passionately for immortality, and as he could not believe wholly and implicitly that CHRIST could or would raise him up from the dead in His own good time, he appealed to the magician to do this for him.

As an answer to this appeal the little work the LEFÂFA ṢEDEḲ, or "Bandlet of Righteousness," was composed by someone who was skilful in fusing Christianity with paganism in such a way that the wayfaring man, whether a fool or not, was led to believe that the composition was a Christian work. And it obtained a great vogue, because it was held to be the most powerful collection of magical texts then known.

The author of the work and the date of its first appearance are alike unknown. There is no evidence that it was translated from the Arabic, though it is possible that there may have been a similar work in Coptic, and it seems that we may regard it as a native production. In its present form it is probably not older than the sixteenth century, and the older of the two manuscripts of it published herein dates from the end of the seventeenth or beginning of the eighteenth century. But the form of the magical element in it is many centuries older, and both it and the beliefs expressed in it certainly were derived from a people who possessed a higher civilisation than that of the ETHIOPIANS, and a superior religion, and elaborate funerary rites and ceremonies. This people I believe to have been the EGYPTIANS, who, even after they had embraced Christianity, mummified their dead, and relied on the efficacy of amulets and spells to effect the preservation and resurrection of the body and to secure for their souls acquittal in the Hall of Judgment and everlasting life, either in the Kingdom of Osiris or in the "Boat of Millions of Years" of the Sun-god.

The Ethiopian, like the Egyptian, attached supreme importance to the knowledge of the secret names by means of which celestial beings lived, for he regarded the name as the vital essence or soul, of every being, whether god or man. In short, the LEFÂFA ṢEDEḲ was regarded as an invincible amulet because it was believed to contain the secret names by means of which the Persons of the Trinity, and Their servants in heaven and upon earth subsisted. Among the magical names given we find the names of all the letters of the Hebrew Alphabet, and, curiously enough, we have the five words of the old and well-known palindrome

SATOR AREPO TENET OPERA ROTAS

turned into the magical names of the five (*sic*) nails by which CHRIST was nailed to the CROSS (see p. 78). This palindrome is also found in a magical text in a Coptic manuscript written in the sixth or eighth century, where it reads

ⲤⲀⲦⲰⲢ ⲀⲢⲈⲦⲰ ⲦⲈⲚⲈⲦ ⲰⲦⲈⲢⲀ ⲢⲞⲦⲀⲤ

(see W. E. CRUM, *Catalogue of the Coptic MSS. in the British Museum*, London, 1905, No. 524, p. 254, col. 2, § vii). We may assume, then, that the Ethiopians borrowed it from the Copts, and it is clear that neither people knew what the words really meant.

The use of the palindrome in magical texts is very ancient and one example at least can be traced back to the third century, viz.

ΑΒΛΑΝΑΘΑΝΑΛΒΑ.

The Gnostics used it frequently, and it is found, in a more or less abbreviated form, on many Gnostic amulets, where, like ADÔNÂY, ABRASAX, ṢABÂÔTH and SEMES EILAM, it appears as a title of the

Pantheus, whether ιαω, YÂH (JÂH) or HORUS, or HARPOKRATES. (See the Gnostic amulets Nos. 60 and 69 in the British Museum, and KING, *The Gnostics*, Plate F, No. 5.) It occurs in Greek magical papyri of the third century (KENYON, *Greek Papyri in the British Museum*, London, 1893, p. 94, line 311), and of the fourth century (*ibid.*, p. 67, line 63), where it is sometimes abbreviated, *e.g.* *αβλαναθ* (*ibid.*, p. 105), *αβλαθ* (*ibid.*, p. 118). In order to obtain the fullest benefit possible from the use of the palindrome ΑΒΛΑΝΑΘΑΝΑΛΒΑ it was necessary to write it in the form in which SERENUS SAMMONICUS (third century) ordered the word of power ABRACADABRA to be written; thus we have :—

1. ΑΒΛΑΝΑΘΑΝΑΛΒΑ	1. ABRACADABRA
ΒΛΑΝΑΘΑΝΑΛΒΑ	BRACADABRA
ΛΑΝΑΘΑΝΑΛΒΑ	RACADABRA
ΑΝΑΘΑΝΑΛΒΑ	ACADABRA
ΝΑΘΑΝΑΛΒΑ	CADABRA
ΑΘΑΝΑΛΒΑ	ADABRA
ΘΑΝΑΛΒΑ	DABRA
ΑΝΑΛΒΑ	ABRA
ΝΑΛΒΑ	BRA
ΑΛΒΑ	RA
ΛΒΑ	A
ΒΑ	
Α	
2. ΑΒΛΑΝΑΘΑΝΑΛΒΑ	2. ABRACADABRA
ΒΛΑΝΑΘΑΝΑΛΒ	BRACADABR
ΛΑΝΑΘΑΝΑΛ	RACADAB
ΑΝΑΘΑΝΑ	ACADA
ΝΑΘΑΝ	CAD
ΑΘΑ	A
Θ	

Now, the Gnostics had both the forms of the palindrome ABLANATHANALBA engraved upon their amulets in the second century, and it seems that SERENUS borrowed the idea of writing ABRACADABRA in the forms shown above from them.

Among the ceremonies to be performed in connection with the use of the book LEFÂFA ṢEDEḲ is the making of the sign of the seal of SOLOMON thrice (*i.e.* once for each Person of the Trinity) over the bier of the dead man with the book. The traditions extant concerning SOLOMON'S seal are somewhat contradictory, and it is clear that some of their writers had confused ideas on the subject. That SOLOMON had a gold ring, with which he worked miracles is generally admitted. This ring contained a bezel made of a magical stone, which was engraved, according to some, with the ineffable name of GOD, YHWH, but whether it was in " square " Hebrew characters, or in letters similar to those found on the Moabite Stone, cannot be decided. Others say that this name was engraved within the Solomonic Pentacle ✯ and others assert that it was inside the Hexagon ✡, similar to that which mediæval astrologers used in connection with ABRACADABRA. JOSEPHUS in his *Antiquities of the Jews* (VIII. 2, § 5) says that the ring with which SOLOMON drove out devils contained a certain *root*, which, as the king is stated to have been an expert herbalist, we may assume was known to possess magical, and perhaps medicinal, powers. In a very interesting Syriac manuscript edited and translated by Prof. H. GOLLANCZ, it is said that twenty-nine magical names were written on [the bezel of] the ring of King SOLOMON, and a list of them is given (*Book of Protection*, London, 1912, pp. 1 and 26).[1] According to the Ḳur'ân (Sûrah XXXVIII) SOLOMON entrusted his ring to one of his concubines called AMINA when he bathed, and one day SAḲHAR, a devil, took the king's form and went to her and took it from her. SAḲHAR went into JERUSALEM, and seated himself on SOLOMON's throne and reigned for forty days; at the end of this period SOLOMON was forgiven by GOD, and the devil fled, and threw the ring into the sea as he went. And a fish swallowed it. A fisherman caught the fish and gave it to SOLOMON, who opened it and found his ring. By means of it he recovered his kingdom, and having caught SAḲHAR with the help of the ring, he tied a stone to his neck and cast him into the LAKE OF TIBERIAS. At a later period, it is said, the ring was taken to JERUSALEM, for it escaped the plundering of NEBUCHADNEZZAR, and was laid up in the Ark of the Covenant in the Holy of Holies, where it remained until the time of TITUS (quoted by Gollancz from The TALISMAN). Abyssinian writers tell a different story. According to them SOLOMON gave his ring to the Queen of SHEBA when she was setting out for her own country, and it was taken back by her son MENYELEK I when he went to JERUSALEM,

[1] In Codex A, p. 54, is given a drawing of the Seal of SOLOMON. In the centre is what seems to be a gem emitting eight rays of light, and between double concentric circles are written the names ḤLYPT SLYT SPILT TR(?)YKT PP MRYT ḤLPT A(?)YLPT. Outside the circles are the names DMPṢ BRWLḤT HKIKT TRKLT PPT PRISHT ALILT PPASHNT SHRI'T PLISHT. The text continues, "These names shall be supporters, and protectors, and deliverers, and protectors (*sic*) against all diseases and sicknesses now; also before" [kings and governors, etc.].

and he showed it to Solomon as a proof that he was his son. Drawings of the Seal of Solomon are found in many Ethiopic amulets, and they are claimed to be copies of the device which was engraved on the bezel of Solomon's ring. A prominent feature in all these drawings is a modified form of the Coptic Cross, which, of course, proclaims their non-Hebrew origin. Worked into the designs are two, four, or eight eyes, which indicate that the Seal was specially intended to protect the wearers of the amulets from the Evil Eye and from the attacks of fiends and the Devil.

The Ethiopic magical texts also say that Solomon used to catch the devils in a net. The following is a tracing made from a rare drawing of the net which is found on an amulet in the writer's possession. Here, too, a form of the Coptic Cross is the most prominent feature in the design. One of the early Christian ascetics commemorated by Palladius held the view that the Devil caught human souls with a

ወርበብተ : ሰሎሞን :

The Net of Solomon.

net, and in the Egyptian *Book of Gates* several of the beings who are going to fight against the enemies of the Sun-god are armed with nets. And it will be remembered that the Babylonian Sun-god MARDUK caught the monster TIÂMAT in a net :—

"He made a net wherewith to enclose TIÂMAT.
"He held the net close to his side, the gift of his father ANU.
"The Lord cast his net and made it enclose her."
(*Fourth Creation Tablet*, lines 41, 44 and 95.)

The general character of the book LEFÂFA ṢEDEḲ was first pointed out by the late lamented scholar, R. TURAEV, who, according to BEZOLD, translated portions of it into Russian, and published them in the *Denkmäler der äthiopischen Literatur*, VII, St. Petersburg, 1908. As I cannot read Russian, and have failed to obtain a copy of his work, I am unable to say how far he carried his researches into the original text. But I feel that I am correct in saying that the Gĕʻĕz (Ethiopic) text and the English translation printed herein are published for the first time. The numbers of the folios in the translation are those of the folios of the manuscript, whilst those given in the photo-lithographic reproduction begin only with the first folio of the text.

My thanks are due to the Trustees of the British Museum for permission to photograph the MSS. Oriental No. 551 and Add. No. 16204; to Dr. Lionel Barnett, Keeper of the Oriental MSS. in the British Museum for facilities in consulting various manuscripts; and to Mr. A. I. Ellis, M.A., F.S.A., Assistant Keeper in the Department of Printed Books, whose knowledge of the contents of our great National Library rivals that of the late Dr. Garnett, for much prompt and time-saving assistance.

E. A. WALLIS BUDGE.

48, *Bloomsbury Street,*
Bedford Square, W.C. 1.
February 17*th*, 1929.

CONTENTS

THE "BANDLET OF RIGHTEOUSNESS"

CHAPTER I

Ethiopian Magical Names of God and their Creative Powers.

Of all the magical works written in Ethiopic and Amharic which have come down to us, the most curious and the most interesting from an archæological point of view is the little book of Lefâfa Ṣedeḳ, which title I have translated by "Bandlet of Righteousness." Very few manuscripts of the work are known, and the only two available to me, viz. those in the British Museum, are reproduced in facsimile at the end of the present volume. The "Bandlet of Righteousness" referred to in the title was a strip of linen or parchment which was exactly as long as the body of the person for whose benefit it was prepared was high, and on this were inscribed a series of eight magical compositions, and, presumably, drâwings of crosses. The width of the strip is unknown; it may have been wide enough to cover the body, but it is more likely that it was only from 3 inches to 6 inches wide, like the linen strips inscribed in hieratic with texts from the Book of the Dead, which the Egyptians buried with their dead in the Saïte and Ptolemaïc periods. This Bandlet was wound round the body of the deceased on the day of burial, and was believed to protect it from the attacks of devils, and enable him to pass through the earth without being stopped at any of the gates or doors, and ultimately to pass into heaven. The possession of this Bandlet ensured for him acquittal in the Judgment, and therefore escape from the awful River of Fire. In fact the LEFÂFA ṢEDEḲ contains in a much-abbreviated and succinct form all the essential elements of the BOOK OF THE DEAD as found in the Recension which was in use in EGYPT during the Græco-Roman period. On these elements are superimposed ideas derived from the writings of the Christian GNOSTICS, and from apocryphal Hebrew works which, probably, in Greek or Syriac translations, were read by the early Egyptian Christians, and from original works in Coptic.

But the peculiar character which the LEFÂFA ṢEDEḲ possesses was given to it by the Abyssinian Christians, who were able to combine the cult of magic with the cult of the VIRGIN MARY. When the ABYSSINIANS adopted Christianity in the first half of the fourth century of our era, theoretically they accepted the doctrine of the Christian Resurrection, and all that it implied. But for centuries they had been believers in native magic, and by its means they attempted to secure for themselves the best things on earth and also everlasting life and the happiness of heaven. They acknowledged that GOD had created the heavens and the earth, and they realized that He was self-subsistent and eternal, but they wanted to find out how He maintained His life and power undiminished, and what was the secret of His being. They believed that if they could only find out this secret they would become as great and mighty as He is. GOD, they believed, had invented magic and given it to them so that they might command the powers of

Nature, and bend them to their will, but with this they were not satisfied, they wanted to be equals of God.

According to a very ancient tradition, which is reproduced in the *Book of the Mysteries of Heaven and Earth*,[1] the Three Persons of the Trinity existed in the waters of the great primeval ocean, and they had had their abode therein for ever. But they existed in *name* only and not in Person. Each Person only assumed His subsequent form by pronouncing His own name. This the Abyssinian theologians interpreted as meaning that each Person possessed a name which at will He could employ as a "word of power." And according to another tradition God, and Michael and all the angels, would have suffered final defeat at the hands of Satan if Michael had not been able to hold up before the rebels a cross of light on which was inscribed the words, "In the Name of the Father, and the Son, and the Holy Ghost." As soon as Satan and his devils saw these words they turned and fled.

Now, these same theologians argued, God not only created Himself by uttering His own name, but the heavens and the earth also, and they came to the conclusion that the Name of God was the Essence of God, that it was not only the source of His power but also the seat of His very Life, and was to all intents and purposes His soul. There is no reason for thinking that they invented this belief concerning the secret Name of God, for the Egyptians had formulated it many centuries before the Abyssinians became a nation. This is proved by a passage in the papyrus of Nesi-Àmsu in the British Museum, in

[1] Edited from a unique Ethiopic MS. in Paris (Bibl. Nat. 117) by Perruchon. (No date.)

which the god NEB-ER-DJER says: "I am he who came into being in the form of the god KHEPERA. I am the creator of everything which came into being. The things which I created, and which came forth from my mouth after I myself had come into being, were many. Heaven did not exist, earth did not exist, and the children of the earth (*i.e.* trees, plants, etc.), and creeping things were not then made. I myself raised them up out of NU (*i.e.* the primeval World-Ocean), out of a state of helpless inertness. I found no place on which to stand. I worked a charm (*i.e.* used a magical formula) upon my heart. I laid the foundations by MAĀT,[1] I made everything that hath form. I was ONE (*i.e.* there was no other), for I had not then sent forth from myself the god SHU and the goddess TEFNUT, and there was none who worked with me. I brought my name [into] my mouth as *ḥeka*, i.e. magic, and I came into being in the form of things that are, and under the form of KHEPERA. I it was who emitted SHU. I it was who emitted TEFNUT. From being the ONE [god] I became THREE [gods]. Plants and trees and creeping things [sprang up] from the god REM. I cried with my EYE (*i.e.* the Sun) and men and women came into being from the tears which fell therefrom."[2]

Among many ancient peoples the utterance of the name was regarded as an act of creation, and the obliteration of a name was equivalent to the destruc-

[1] MAĀT, the personification of physical and moral law and order. The part which she played at the creation resembles that of "Wisdom" which is described in Proverbs viii. 23 f.

[2] See my hieroglyphic transcript of Papyrus, No. 10,188, with a transliteration and translation in *Archæologia*, Vol. LII, London, 1891. A facsimile of the hieratic text is published in my *Egyptian Hieratic Papyrus in the British Museum*, Vol. I, London, 1914 folio.

tion of the person who bore it. The EGYPTIANS thought that any abuse of a man's name injured him personally, and when standing in the Hall of Judgment before OSIRIS the deceased prayed fervently that his "name might not be made to stink" in the presence of the Assessors of the Great God. Several compositions were written by the priests with the special object of making a man's name to "germinate," *i.e.* to flourish and not to be forgotten. On tombs, stelæ, papyri, amulets and every object buried with the dead in their tombs, the names of the deceased are repeated *ad nauseam*, for how could a nameless soul be presented to OSIRIS? One of the chief objects of the funerary spells which were written by the Egyptians was to supply the dead with the names of the various beings, and gates and doors, and their guardians, which they would meet within the Ṭuat. By the use of these the deceased was able to say, when he entered the Hall of Judgment, "O Great GOD, I have come to thee, O my Lord, and I have brought myself hither that I may behold thy beneficence (or beauties). I know thee. I know thy name. I know the names of the two-and-forty gods who are with thee in this HALL OF MAĀTI, who live as wardens of sinners, and who feed upon their blood on the day when the lives of men are reckoned up in the presence of the god UN-NEFER" (*Book of the Dead*, Chap. CXXV).

The knowledge of the name of a god enabled a man not only to free himself from the power of that god, but to use that name as a means of obtaining what he himself wanted without considering the god's will. And from the words of St. John (Rev. ii. 17) it may be gathered that Christians regarded the gift of a white stone inscribed with a new name, which no man except the recipient knew, as one of the greatest gifts which GOD could bestow on a servant of His. In primitive times the name of the king was regarded with reverence such as was due to a god, and his subjects had it engraved on their rings, seals, and scarabs, believing it to be a protection for them; and there is little doubt that it was used by many as a word of power. We see in inscriptions that it is enclosed within an oval, now called "cartouche," having a bar at one end of it. The line of the oval and the bar represent a rope, the two ends of which are tied in a knot, and they were supposed to give magical protection to the royal name. The weaving of magical knots was a well-known art among ancient magicians, and it is practised by EGYPTIANS and ARABS at the present day.

Returning now to the LEFÂFA ṢEDEḲ, we see that the person to whom we really owe GOD's revelation of His secret name is the VIRGIN MARY. Her grief and tears and sorrow for the sufferings which she imagined her kinsfolk would be forced to undergo in the LAKE OR RIVER OF FIRE won the compassion and help of her Son, the WORD; and He did not rest until GOD the FATHER had dictated to Him the secret and magical names in the Book which He had composed before CHRIST was born in the flesh. That these names were numerous need not surprise us, because they are only descriptions of GOD's own attributes, and aspects, and powers. From the seventeenth chapter of the BOOK OF THE DEAD we know that all the gods of all the great companies of gods were only the names of the attributes and powers of the great Sun-god, whither he was called KHEPERA.

NEB-ER-DJER, TEM, RĀ OR ÅMEN. In the great LITANY OF RĀ praises are rendered to the Seventy-five chief forms of RĀ, each of whom has a distinct name; and in the LITANY OF OSIRIS, which is found in Chap. XV of the BOOK OF THE DEAD, we have addresses to the Nine forms of OSIRIS, and each form has its proper name. In a Demotic papyrus we have a whole string of names of the god whose name was formed of the vowels of the Greek alphabet—"Iao, Iaolo, Therentho, Psikhimeakelo, Blakhanspla, Iac, Ouebai, Barbaraithou, Ieou, Arponknouph, Brintatenophri, Hea, Karrhe, Balmenthre, Menebareiakhukh, Ia, Khukh, Brinskulma, Arouzarba, Mesekhriph, Niptoumikh, Maorkharam." And again, "Laankhukh, Omph, Brimbainouioth, Segenbai, Khooukhe, Laikham, Armioouth" (GRIFFITH, *Demot. Mag. Pap.*, p. 111). Similarly in the Coptic BOOK OF IEU we have a long series of lists of the names of the emanations of the god IEU = ΙΑΩ = JÂH. Many of these were cut upon stones as charms, and those who were instructed knew that they were the names of the forms and attributes of the GREAT GOD.

The GNOSTICS followed the example of the ancient EGYPTIANS, and their spells consist usually of a string of names of the Æons, the head and chief of whom is GOD. Here is an instance:

ΑΤΩϹΑϹΑΩΑΛΩΝΕ
ϹΕΜЄϹЄΙΛΑΜΑΒΡΑϹΑΞ
ϹΞΖΥΡΡΑΤΗΑΚΡΑΜΜΑ
ΚΡΑΜΜΑΚΑΝΑΡΙϹϹϹЄ

Here in the first two lines we recognize the names ALÔN, SEMES EILAM, and ABRASAX, and the remainder of the inscription no doubt contains many others.

In dealing with inscriptions of this kind we must always remember that both the GNOSTICS and the COPTS believed that our Lord spoke to MARY, and that she replied to him in a language which was known only to themselves. Thus CHRIST addressed MARY in these words: MARI KHAR MARIATH, *i.e.* "MARY, mother of the Son of GOD," and MARY replied, ḤRAMBOUNE KATHIATHARI MIÔTH, *i.e.* "The son of the Almighty, the Master, and my Son" (BUDGE, *Coptic Apocrypha*, p. 189). It is possible that some of the spells in the LEFÂFA ṢEDEḲ may be transcripts from Coptic originals. Examples of the language which CHRIST used in speaking to His Father are given in the PISTIS SOPHIA, *e.g.* AELIOUÔ, IAÔ, AÔI, ÔIA, PSINÔTHER, THERNÔPS, NÔPSITER, ZAGOURI, RAGOURI, NETHMOMAÔTH, NEPSIOMÂTH, MARAKHAKHTHA, THÔBARRABAÔTH, THARNAKHAKHAN, ZOROKOTHORA, IEOU, SABAÔTH (KING, *Gnostics*, p. 285; AMÉLINEAU, *Pistis Sophia*, p. 185). And the GNOSTICS believed that CHRIST revealed to the disciples the names of the Aeons who forgave sins, viz. GIPHIREPSINIKHIEOU, ZENEI, BERIMOU, SOKHABRIKHIR, EUTHARI, NANAÏDIEISBALMIRICH, MEUNIPOS, KHIRIE, ENTAÏR, MOUTHIOUR, SMOUR, PEUKHIR, OOUSKHOUS, MINIONOR and ISOKHOBORTHA, and the names of the "Great Powers," viz. AOUIR, BEBRÔ, ATHRONI, IOUREPH, IOVE, SOUPHEN, KNITOÛSOKHREÔPH, MAOUÔNBI, MENEUÔR, SOSÔNI, KHÔKHETEÔPH, KHÔKHE, ETEÔPH, MEMÔKH and ANIMPH.

Like the EGYPTIANS,[1] GNOSTICS and COPTS the

[1] The god MARDUK also possessed a large number of names; according to the Creation Legend the gods proclaimed his Fifty Names fifty times. See *Babylonian Legends of the Creation* (British Museum), p. 65.

MUḤAMMADANS possessed a long series of names of ALLÂH, and lists of them were written on amulets and talismans and worn by men and women alike as protectors of their souls and bodies. The great and essential name of GOD according to Muslim writers is ALLÂH, which is known as "Ismu az-Zât," *i.e.* the essence name; all the other names of GOD, including "AR-RABB," are regarded as the "Asmâ'u aṣ-Ṣifât, *i.e.* "names of the attributes." MUḤAMMAD, the Prophet, says in his Ḳur'ân (Sûrah vii, l. 179) that GOD has a number of "beautiful names" (al-Asmâ'u al-ḥusnā), and that they are Ninety-nine in number, and that whosoever reciteth them shall enter into Paradise. The following are specimens of these names:

Ar-Raḥmân	The Merciful	Al-Khâliḳ	The Creator
Ar-Raḥim	The Compassionate	Al-Bârî	The Maker
Al-Malik	The King	Al-Muṣawwir	The Fashioner
Al-Ḳuddûs	The Holy One	Al-Ghaffâr	The Forgiver
As-Salâm	The Peace	Al-Ḳahhâr	The Dominant
Al-Mu'min	The Faithful	Al-Wahhâb	The Bestower
Al-Muhaimin	The Protector	Ar-Razzaḳ	The Provider
Al-'Azîz	The Mighty	Al-Fattâh	The Opener
Al-Jabbâr	The Repairer	Al-'Alîm	The Knower
Al-Mutakabbir	The Great	Al-Ḳâbiẓ	The Restrainer

Thus we see that the ABYSSINIANS, like the pagan EGYPTIANS, and the Christian EGYPTIANS, *i.e.*, COPTS, and the Gnostic sects who based their magical systems chiefly upon African cults, assigned to GOD a whole series of magical names which they used as words of power. All these peoples ascribed to the name of GOD or of a man an importance which it is impossible for us to realize fully because we do not know the exact meaning which they attached to their words for "name." It is clear, however, that they believed that the life and existence of a god or a man were bound up with the existence of his name inextricably; neither god nor man could exist without his name, and the "killing" or destruction of his name was equivalent to the destruction of his existence. Mr. EDWARD CLODD thinks that the Celts and perhaps the whole Aryan family believed that the name was not only a part of a man, but that it was that part of him which is termed the soul, or the breath of life (*Magic in Names*, p. 280). He is undoubtedly correct as far as the peoples he mentions are concerned, and the evidence supplied by Egyptian, Coptic, Gnostic, Hebrew, Arabic and Ethiopic texts convinces me that the same may be said of the AFRICANS and SEMITES.

We have seen that the LEFÂFA ṢEDEḲ was believed to secure for the dead the preservation of their bodies, and life beyond the grave, and entrance into heaven, but nothing is said in it as to means which the dead are to employ for the maintenance of their life whilst proceeding to heaven. The EGYPTIANS in their Books of the Dead supplied the deceased with magical formulas which, when recited by him, produced clean water, bread-cakes, roast meats, clean linen apparel, unguents and perfumes, etc., and to make sure that he should lack nothing in the Ṭuat they made offerings frequently in his tomb. The kinsmen of the deceased, or a priest, separated the spiritual parts of these offerings by means of spells, and thus the life of the KA was maintained. Now of the funerary rites and ceremonies of the ABYSSINIANS, certainly in the old times, practically nothing is known. Men and women belonging to classes of no social importance were carried out of their houses as soon as they died, probably on the mats they died on, to the edge of the village or town, and laid in shallow trenches, without any further ceremony. Stones were laid on and about the body with the view of keeping the jackals, wolves, and foxes from devouring it, but they rarely prevented these animals from obtaining their nightly meals on human flesh. The ABYSSINIANS generally pay little respect to the dead, unless they happen to be kings or members of the Royal Family and high ecclesiastical officials. Mr. C. F. REY, the distinguished traveller, tells us that " in Addis Ababa the principal place of burial near the market place is ridden and walked over by passers, and occasionally at night the jackals and hyenas come up from the river, dig up the lightly covered remains and indulge in gruesome banquets " (*Unconquered Abyssinia*, p. 79). But the matter is very different when the deceased is a person of high rank and position. Then the body is washed and rubbed with unguents, and wrapped in cloth or Indian and Persian silks, and priests chant the penitential Psalms and recite prayers and burn incense. In due course the body is taken to the church in which it is to be buried, accompanied by a crowd of priests and soldiers, and a great mob of the ordinary people.

As far as I know no modern traveller has described the funeral service of an Abyssinian royal personage, but we know from the manuscripts preserved in London, Oxford and Paris what the general character of the contents of the MAṢḤAFA GENZAT, *i.e.* the Book of Burial, is. It opens with a series of miscellaneous prayers that mercy may be shown to the dead, and these are followed by the recital of the prayer of ST. ATHANASIUS for the passing of the soul. Then follow : the Absolution in one of its three forms, the penitential prayers given by GOD to ST. PETER, prayers composed by the VIRGIN MARY, an admonition which is said for every dead person, a funeral sermon by JACOB OF SERÛGH or by 'ABBÂ SALÂMÂ, the recital of some narrative in which the importance of giving alms is inculcated, and then Benedictions by Fathers of the Church. (See Brit. Mus., MS. Add. 16,194; Oriental MSS. 551–555.) The ABYSSINIANS even to this day seem to have no special Mass for the Dead, but they have hundreds of prayers for the dead, and they commemorate the dead frequently and make offerings and burn incense to them, and it is believed that the prayers which are made whilst incense is being burned are carried up to heaven on the smoke thereof. The following is the translation of a funerary prayer:

"O Lord, remember our fathers and our brothers who have died in the True Faith, and do Thou make their souls to enjoy rest with the saints and the righteous. Lead them on their way and gather Thou them together in a place of well-being, where living water is to be found, and in a Garden of Delights." It was a heaven of this kind for which the Egyptian also prayed. In the same Anaphora, which is attributed to Saint BASIL, a prayer is made on behalf of those who pray for the dead, and who make offerings to them. (See JEROME LOBO, *Voyage Historique*, Paris, 1728, p. 345.) According to some, the offerings are intended as payment to the priests for their services, but others say that the dead are benefited by them, and that they help them to escape from places of torture in the Other World. F. BALTHAZAR TELLEZ says that the dead are bewailed for many days together, and that the lamentations are continued throughout the day. The mourners beat drums, clap their hands, smite their breasts and faces,'' uttering such dismal expressions, in a doleful tone, that they torment the head and grieve the heart.'' When a cavalry soldier is buried, his horse, spear, shield, clothes and weapons are taken with him to the grave (*Travels of the Jesuits in Ethiopia*, London, 1710, p. 44).

CHAPTER II

Description of the Manuscript and its Contents.

The manuscript (A) containing the Ethiopic text of the Lefâfa Ṣedeḳ which is translated in the present work is preserved in the British Museum, where it bears thc number Add. 16,204.[1] It was presented to the Trustees by the Church Missionary Society on 20th August, 1846, and was brought from Abyssinia by one of their missionaries, either Dr. J. Lewis Krapf or Dr. C. W. Isenberg, most probably the latter. The manuscript was briefly described by Dillmann, who says that its size is octavo, that it contains 30 parchment leaves, and that the well-written text on each page is arranged in two columns (*Catalogus Codd., MSS. Orientalium qui in Museo Britannico asservantur*, London, 1847, No. LXXIX, p. 64). There are two compositions in the manuscript, viz. :

1. Lefâfa Ṣedeḳ, ለፋፈ ፡ ጽድቅ ፡ which contains the series of eight magical spells that form the "Book of Life," *Maṣḥafa Ḥaywat* መጽሐፈ ፡ ሕይወት ፡

[1] The manuscript from which the second version of the Lefâfa Ṣedeḳ is taken (MS. B) is a fine volume measuring 12⅛ in. by 10¾ in. and containing 151 folios. It was written in the latter half of the eighteenth century and is numbered Oriental 551. A full description of the contents of this manuscript will be found in Wright, *Catalogue of the Ethiopic Manuscripts in the British Museum*, London, 1877, No. CXLIV, p. 98.

These are written in Ethiopic, and are accompanied by figures of the Cross of an unusual character. Fol. 2*a*–26*b*.

2. Maṣḥafa Terguâmê̤ Fîdal, መጽሐፈ፡ ትርጓሜ፡ ፊደል፡ a short work, written in Amharic and dealing with the names of the Persons of the Trinity and theological expressions. Fol. 27*a*–30*a*.

The manuscript measures 7 in. by 5 in., and is written in a good clear hand, probably of the first half of the seventeenth century.

On Fol. 1*b* is the following description of the contents of the Lefâfa Ṣedeḳ, probably in the handwriting of the great Amharic scholar and missionary, Dr. C. W. Isenberg: "Lefafa Ts'edk, *i.e.* Supplication of Righteousness—one of the most striking pieces of Abyssinian absurdity and superstition. The names of Christ, real and invented, some of them shocking (*e.g.* Satanael, etc.), used as a spell against unclean spirits, against all evil, and death." But the good and zealous missionary did not realize that in this little work clues to the primitive, fundamental beliefs of the Abyssinians are to be found, and that it was to the Abyssinians precisely what the Book of the Dead was to the ancient Egyptians. The spells and stories contained in it may by some be regarded as absurd, incredible and impossible, but they are only the necessary outcome of a long series of ancient beliefs which have been current in Abyssinia, and in other countries of north-east Africa, from time immemorial, and which the teachings of Christianity in them for fifteen hundred years have not yet been able to eradicate. Of the way in which the book Lefâfa Ṣedeḳ was used Isenberg says nothing, but Dr. Krapf, having described it as containing "prayers and exorcisms against evil spirits," goes on to say that it is "a book much prized by the ABYSSINIANS, and often buried with their dead" (*Travels and Missionary Labours in East Africa*, London, 1860, p. 556, No. 39).

DILLMANN carries us a step further, for he says that "Lefâfa Ṣedeḳ," is the name given to the strips of parchment inscribed with magical prayers which are wrapped round the bodies of the dead and are buried with them in their graves. He then goes on to say that in their mad superstition the modern ABYSSINIANS believe that men who are provided with such inscribed strips of parchment will come forth from the Judgment before God uncondemned.[1] The *tæniæ membranaceæ* to which he refers are undoubtedly the little strips of parchment inscribed with versions of the fight between Saint SÛSENYÔS (SISINNIOS) and the arch-devil WERZELYÂ,[2] ውርዜልያ፡ (in Coptic ⲃⲉⲣⲍⲉⲗⲓⲁ), and with short spells against sicknesses and diseases, and magical figures and crosses, and names which exist by the hundred in the great Libraries and Museums in EUROPE. But these form a class of documents by themselves, and have very little in common with the LEFÂFA ṢEDEḲ, either as regards origin or final purpose.

[1] homines talibus fasciis ornatos coram Deo justificatum iri vesana Abyssinorum recentioris ætatis superstitio imaginatur (*Lexicon*, col. 66).

[2] For translations of the legend see FRIES in the *Actes* of the VIIIth Oriental Congress, Leyden, 1893, pp. 55–70; BASSET, *Les Légendes de S. Têrtâg et de Susenyos*, Paris, 1894; WORRELL, *Studien zum abersinischen Zauberwesen*, in *Zeit. für Assyriologie*, Bd. XXIII. p. 168. In succeeding volumes of the *Zeitschrift* the last named scholar has published descriptions and translations of series of such parchment rolls.

The little rolls of inscribed parchment or paper to which DILLMANN refers were written by scribes for men and women to wear as amulets, and in none of them is, so far as I have seen, the claim made that the compositions are of divine origin. On the other hand, it is distinctly stated in the LEFÂFA ṢEDEḲ that the work was written by GOD Himself, and copied by our LORD with a pen of gold, and that the names revealed in it are those by which the FATHER, and the SON, and the HOLY GHOST maintained their existence, and governed the heavens and the earth and all that is in them. The parchment amulets contain prayers or spells, the recital of which was supposed to preserve men and women from sicknesses of the body of every kind, and to save women from miscarriage and abortions caused by evil-disposed devils, and to ensure their safe delivery. But the LEFÂFA ṢEDEḲ was written with the special object of preserving the bodies of the dead from mutilation, and from the attacks of devils, and from the awful River of Fire in hell, and enabling their souls to attain to everlasting life and health and well-being in the kingdom of heaven. It was for this reason that the ABYSSINIANS, as Dr. KRAPF tells us, buried copies of it with the dead. When they first began to do this cannot be said, but the conservatism of the ABYSSINIANS in all matters connected with the burial of the dead has always been so strong, that we are justified in assuming that the custom of burying copies of the LEFÂFA ṢEDEḲ with the dead has been in existence for several centuries. In any case we are entitled to call that work an ETHIOPIAN BOOK OF THE DEAD. There is no proof that the ABYSSINIANS borrowed the custom from the ARABS, who never have buried, and still do not bury, holy books or amulets with their dead. It is far more likely that the ABYSSINIANS borrowed the custom from the ancient EGYPTIANS or the COPTS. It is unlikely that the custom was universal at any time, for to the poor the cost of the parchment and the fee of the scribe would naturally be prohibitive.

The temples at JABAL BARKAL, and many of the ruined buildings at NAPATA, show that the native kings of NAPATA, in the centuries immediately preceding the Christian era, brought workmen and funerary masons to repair the ancient buildings, and cut or recut on them hieroglyphic inscriptions. The chapels of the pyramids of MEROË are decorated with reliefs and paintings containing series of vignettes from the Saïte Recension of the BOOK OF THE DEAD. And the hieroglyphic funerary texts cut on their lintels and door jambs and walls are manifestly the work of skilled Egyptian and not Nubian workmen. From these the natives of MEROË would learn much concerning the Egyptian belief in the efficacy of magical funerary spells, and the masters and men of caravans trading with NORTHERN ABYSSINIA would carry stories of what they had seen to their kinsfolk and neighbours generally in their native land. In EGYPT, MEROË was regarded as the home of "black magic," and of the spells which were employed in connection with the dead. The knowledge of the "black art" of EGYPT entered ABYSSINIA by two channels, viz. by way of the Blue Nile, and by way of the caravan road which the merchants of ADULIS used when sending their merchandise to AKSÛM. BRUCE relates (*Travels*, ii. p. 35) that his friend the king of ABYSSINIA brought back to GONDAR from

TIGRAY a black stone "cippus of Horus," 14 inches high and 6 inches wide.[1] On one side were sculptured figures of the gods of EGYPT, and on the back and edges were copies of well-known spells in Egyptian hieroglyphs.

A considerable number of antiquities of this class are known, and good examples are to be seen in the British Museum. The EGYPTIANS placed them in their houses and temples in order to protect those who were in them from the attacks of fiends and devils, and noxious animals and reptiles, whether in their natural forms or magical disguises. The cippus found in TIGRAY shows that the knowledge of the use of such objects had penetrated ABYSSINIA at some period between the sixth and first centuries B.C. It must not be forgotten that the people of GESH, *i.e.* NUBIA, and the ISLAND of MEROË, were skilled magicians, and that they claimed to possess the power of making their spells effective in places as far removed from MEROË as THEBES and MEMPHIS. Thus we read in the *Stories of the High Priests of Memphis* (ed. GRIFFITH, Oxford, 1900, p. 179) that HOR, the son of the Negress, made a litter of wax and four bearers, that he read a spell over the figures of the men and breathed into them the breath of life, and ordered them to go to EGYPT and bring its Pharaoh back with them to the Viceroy's palace, where he was to be taken with 500 stripes. Through the operation of Hor's sorceries the figures took their litter to EGYPT and brought Pharaoh back to GESH, and when he had been beaten in the Viceroy's presence with 500 stripes, they took him back to EGYPT, all

[1] It was found at Aksûm in 1771; I have been unable to find out where it is now.

within the space of six hours! The officers of Pharaoh's Court did not believe the story he told them about his transport to Gesh until he showed them his back and the weals which the blows of the stick had raised on it.

CHAPTER III

THE TITLE LEFÂFA ṢEDEḲ, ልፋፈ ፡ ጽድቅ ፡

Dr. KRAPF translated LEFÂFA ṢEDEḲ by " Supplication of Righteousness," but this is manifestly a wrong translation. In his *Catalogue of the Ethiopic MSS. in the British Museum* (p. 61), published in 1847, DILLMANN translates the title by " Volumen Veritatis," and in a note says that the word LEFÂFĔ is not to be found in the Ethiopic or Amharic Dictionaries, and that he is obliged to seek a meaning for it in Arabic. In his great *Lexicon*, published in 1865 (col. 66), he gives as the original of the word the Arabic *lifâfah* لِفَافَة, a noun derived from the root *laffa* لَفَّ, meaning " to wrap up," " to envelop," " to twine," " to bandage," and the like. We must therefore render *lefâfĕ* (in the genitive *lefâfa*) by " bandage," " wrapper," " wrapping," " bandlet," " fillet," " chaplet," or some such word. D'ABBADIE, GUIDI and ARMBRUSTER make no mention of this word, and the only word of a somewhat similar sound which they give is *lĕfâfî*, which means a " tree stripped of its bark." It is well known that the ABYSSINIANS wrapped bodies of their dead in large sheets of cotton or linen, and that when the wrapping up was finished it seemed as if they were placed in bags or sacks [1];

[1] " Il morto vien lavato da capo a piedi e profumato, . . . lo avvolgono in largo lenzuolo come in un sacco." LINCOLN DE CASTRO, *Nella Terra dei Negus*, Vol. I, p. 298.

such a wrapping we might translate by "shroud" without doing violence to the word *lĕfâfĕ*. The second word in the title of the book, ṢEDEḲ, means "truth," "justice," "righteousness," and "justification," and therefore the meaning of LEFÂFA ṢEDEḲ is the "Bandlet of Righteousness," or "Fillet of Justification [in the Judgment [1]]." In the Rubrics in the work it is directed that the book, in whole or in part, is to be attached to the neck of the body, living or dead, a direction which is found in the Rubrics to many of the chapters of the Egyptian BOOK OF THE DEAD. The EGYPTIANS often swathed their dead in sheets or strips of linen or papyrus in which specially selected chapters were written, *e.g.* the mummies of THOTHMES III and ȦMENḤETEP III, and the mummy of ḤENT-MEḤIT, high priestess of ȦMEN; but whether the ABYSSINIANS followed their example and inscribed the shrouds of the dead is not known. It is probable that the texts which were attached to the neck were written on goat-skin or sheep-skin.

[1] Perhaps even "Shroud of Righteousness."

CHAPTER IV

The Lefâfa Ṣedeḳ and the Book of the Dead.

All the sections of the book begin with the words : " In the Name of the Father, and the Son, and the Holy Ghost, One God," and between the end of one section and the beginning of the next is an elaborate figure of the Cross. The figures of the Cross were added to the texts with a view of increasing their potency, and the ideas of magic underlying their introduction here are identical with those of the ancient Egyptian scribes, who added the magical pictures which are now described as " Vignettes " to the various chapters of the Book of the Dead. The Cross gave " life " to all mankind, and the picture of it gave life, both in this world and the next, to the man who read, or caused to be read, or wrote, or caused to be written, or recited, the *Ṣalôtât* or " prayers," *i.e.* magical formulas or spells, found in this book.

The First Section contains the Ṣalôt ba'e[nta] Madkhânît, *i.e.* the " Prayer for redemption (or salvation)," which is taken from the Maṣḥafa Ḥaywat, *i.e.* the " Book of Life," which is called " Lefâfa Ṣedeḳ." This prayer was written by God the Father, with His own hands, before Christ was born of the Virgin Mary. But why should the Father write such a prayer, and to whom was He to address it when written? To this question two answers are possible. He either composed it and wrote it down because He knew that His Son would require it from Him for the use of the children of men, or He composed and wrote it because He Himself had on some occasion been in urgent need of such a prayer. The ABYSSINIANS saw nothing incongruous in assuming that GOD used magic, especially in connection with His secret or hidden name, for His own benefit or in effecting his purposes and designs. This prayer was revealed to CHRIST after His Incarnation, and He transmitted it to the VIRGIN MARY. And the pious ABYSSINIAN argued that if GOD had found the prayer useful, and a means of deliverance from some danger or attack, it was all-important for a man to obtain knowledge of it. The special merits claimed for the prayer in the opening paragraph are:

1. It will make a man to pass through the narrow gate.
2. It will bring him into the kingdom of heaven.
3. It "guideth [to] righteousness," or truth, *i.e.* it is a sure guide (?).

Now, in the Rubrics to some of the chapters of the Egyptian BOOK OF THE DEAD the same benefits are promised to those who use the spells in that Book. Thus in the Rubric to the shorter version of Chapter LXIV, which is said to contain the substance of the whole work, it is said, "[If this Chapter be known] by a man he shall come forth by day, and he shall not be repulsed at any gate of the Ṭuat (Underworld) . . . he shall not die, and behold, the soul of that man shall flourish. . . . It is a great protection [provided by] the god." In the Rubric to Chapter LXXII we read, "If this Chapter be known on earth, or written on the coffin of a man . . . he shall enter into the Ṭuat (Underworld) and not be driven back. . . . He shall enter in peace into SEKHET AARRU[1] . . . and he shall flourish there as he did upon earth." In the Rubric to Chapter XCI it is said that the deceased "shall never be held captive at any door in Amentt" (*i.e.* the kingdom of OSIRIS). In the Rubric to Chapter C, it is said that if a copy of the chapter written on new papyrus be attached to the breast of the deceased, the god THOTH shall number him among the elect, and he shall live with RĀ daily. The Rubric to Chapter CXXXVI A says that the deceased shall have his being "among the living, and he shall never perish; and he shall have an existence like unto that of the holy god; no evil thing whatsoever shall attack him . . . he shall not die a second time . . . he shall live and shall become like unto the god [OSIRIS]."

The Rubric of Chapter CXXXVII A is more explicit. The recital of this chapter would make the deceased "a living soul for ever." In the eyes of the gods, and the AAKHU (*i.e.* beatified Spirits) and the MITU (*i.e.* the dead or the damned?) who were in the Underworld he would assume "the form of the Governor of AMENTT" (*i.e.* OSIRIS) and he would have power and dominion like that god. The deceased would pass without hindrance through the seven halls of heaven, and no limit to his journeyings would be set for him. And the Rubric continues, "He shall never, never, have a sentence of condemnation passed upon him on the days of the Weighing of Words by the god OSIRIS." That is to say, the knowledge and recital of the

[1] "The Field of Reeds," a section of the Egyptian Heaven.

chapters, and the performance of the ceremonies ordered to be performed in connection with it, would make it certain that on the day of the Judgment which took place in the Hall of MAĀTI before the Forty-two Assessors and the god OSIRIS, the deceased would triumph when his soul was weighed in the Great Scales, and he would enter into the kingdom of OSIRIS as one who was "true in word and in deed." And the words of the chapter which produced this result for the deceased were to be copied accurately from the "Book which ḤERṬAṬEF, the son of KHUPU (CHEOPS), discovered in a coffer in the sanctuary of the temple of the goddess (UNNUT, the Lady of UNNU (*i.e.* HERMOPOLIS, or the city of THOTH)."

Now this book was in the "writing of the god himself." The god referred to here is THOTH, who, according to the theologians of MEMPHIS, was the heart of PTAḤ, and according to the most ancient tradition of HELIOPOLIS, the heart of ATEM, or ATUM-RĀ, or RĀ; THOTH was also called the "Tongue of RĀ," and was regarded as the great divine author *par excellence.* But according to texts of the New Kingdom, and of the Saïte period, THOTH was the eldest son of RĀ and the firstborn of the gods, and as such many of the attributes of his father were assigned to him. He is even said to have been "self-born" or "self-produced," and as such he became the creator of the universe. But as BOYLAN observes (*Thoth,* p. 120) he did not fashion gods and men like PTAḤ, the sculptor, or beget them, as did ÀMEN, or make them on a potter's wheel like KHNEMU, but he *thought* them out, being the heart of RĀ or TEM, and being the tongue of the "great god," he "commanded and they were created." The word of THOTH gave being to his thought, and as the speaker of words which gave being to his thoughts, the words and formulas which were uttered by him were believed to possess invincible and magical powers. He was regarded as the author of the spells and prayers in the BOOK OF THE DEAD in all its Recensions, and the EGYPTIANS never doubted that the living and dead who were blessed by the words of THOTH were blessed, and that those cursed by his words were accursed indeed. Thus the Abyssinian Christians assigned to GOD ALMIGHTY the authorship of the magical spells and names in the LEFÂFA ṢEDEḲ, in precisely the same way as the Egyptian scribes attributed to THOTH the authorship of the spells and prayers in the PERT-EM-HRU, *i.e.* the [Book of] Coming forth by Day, or the BOOK OF THE DEAD. And both works were believed to produce the same results when recited, viz. to preserve the bodies of the dead intact and to procure for their souls everlasting life coupled with comfort and happiness.

The words "it guideth [to] righteousness," or truth are probably the work of the scribe who wished to assure the reader of the great value of the spell following.

CHAPTER V

The Contents of the Book of Lefâfa Ṣedeḳ described.

The First Section

How the Lefâfa Ṣedeḳ came to be known upon earth is next described. On the sixteenth day of the month Yakâtît (*i.e.* February), our Lord appeared to Mary His mother, when she was in the Garden (*i.e.* Paradise) inspecting the abodes of the righteous. From the place where she was she was able to see also the various divisions of hell in which those who had been condemned in the Judgment were suffering the punishments which the sins committed by them on earth had brought upon them. Abyssinian tradition states that Mary was conducted through heaven and hell by our Lord, who explained to her why the various grades of the righteous were permitted to enjoy their bliss and happiness, and why the different groups of sinners were made to suffer the horrible tortures which she was able to see with her own eyes. According to a manuscript in the British Museum (Orient. 605, Fol. 94*a* ff.) Mary described her visit to heaven and hell to John, the Son of Zebedee, and the text of her description as reported by him is given in that manuscript. Complete translations of the document and of others of a similar character are given in my *Legends of our Lady Mary*, London, 1922, p. 245 ff. John and other writers of works of the same kind borrowed largely from older apocryphal works, such as the *Apocalypse of Peter* (see M. R. JAMES, *Apocryphal New Testament*, Oxford, 1924, p. 505 ff.), portions of which were translated from Greek into Latin, and later into Coptic and Ethiopic.

The sight of the blessed in the City of God filled MARY with joy and gladness, but when she came to the abode of the damned " she was stupefied with horror, and quaking came upon her, and she feared greatly." Thanks to the *Apocalypse of Peter*, and the *Vision of Heaven and Hell*, which she dictated to her favourite saints, we learn that that which terrified her most was the sight of the tortures of the damned in the RIVER OF FIRE. There she saw some men and women suspended by their tongues or feet over a blazing fire; and others standing in liquid fire (boiling water ?) up to their waists, whilst worms gnawed at their entrails; and others who were being fried in large pans, as fish are fried on earth. In one part of the RIVER OF FIRE were men and women who were being stung by vipers, cobras, and scorpions, whilst their bodies were burning; in another the damned were beating each other with red-hot rods, and stumbling about in the flames to avoid each other's blows; and, worst of all, the tortures were never-ending (for further details, see p. 88 f.). When our Lord saw that MARY was overcome by the sight of the sufferings of the damned He bade her to put away fear and reminded her that she had brought Him forth. In reply MARY asked Him to what purpose had she borne Him if such terrible things continued to happen? How were JOACHIM and ḤANNAH, her father and mother, and her brother and her sister ELISABETH, and even king DAVID, to escape from this awful RIVER OF FIRE? And MARY urged our Lord to tell her, clearly and truly, how they were to be saved. Our Lord replied that He was unable to tell her, for if He did the means would become generally known, and then men would commit sin wilfully, because they knew of a way whereby they might escape the penalties of their sins. On this MARY wept bitterly and again asked Him to what end she had carried Him in her womb for nine months and five days? And seeing her grief, her Son wept in sympathy with her, and promised her that He would converse with the Father on the matter, and that when He had received permission from Him He would tell her how her family was to be saved from the RIVER OF FIRE.

The idea of the RIVER OF FIRE in hell was borrowed from the Egyptians, and there are several allusions to it in the BOOK OF THE DEAD. Thus we read of a monster who watched by the "Bight of the Lake of Fire," and devoured the bodies of the damned that passed him (Chap. XVII, ll. 39–41). The recital of Chapter XVIII, according to its rubric, would enable a man to escape from every fire, and the rubric of Chapter XX says that if a man is purified with water of natron "he shall come forth [in safety] from the fire." Some chapters suggest that the RIVER OF FIRE was really a lake of boiling water (*e.g.* Chapter LXIII), which scalded the wicked who entered it but seemed as cool as dew to the righteous. The Vignette to Chapter CXXVI gives a picture of this lake, and three wavy lines representing water are seen in the middle of it. In one part of the Egyptian hell there were five pits of fire, which seem to be referred to in the *Apocalypse of Peter*, and pictures of these are given in the Book *Ammi Ṭuat*. Each was presided over by a fire-goddess, who supplied the fire from her own body; in the first two pits we see the wicked being consumed, in the third are the souls of the damned, in the fourth their shadows, and in the fifth their heads, see my *Egyptian Heaven and Hell*, Vol. III. p. 249. The "Boiling Lake," is represented in the *Book of Gates* (*ibid.*, Vol. II. p. 108), and the text says, "The water of this lake is boiling hot, and the birds fly away when they see its waters, and smell the fœtid odour thereof."

Then JESUS went to His Father and told Him that MARY was weeping, and asked Him to give Him the MAṢḤAFA ḤAYWAT, *i.e.* "Book of Life," which God had written with His own hand before JESUS was brought forth by MARY. In answer God said that He would give it to Him because He could withhold nothing from MARY, and because it was meet for Him to reveal everything to His Son. Then a cloud of light spread itself over them, and seven veils or pavilions of fire surrounded the Father and the Son, and in the secrecy which these afforded JESUS wrote down with a pen of gold the words of the book LEFÂFA ṢEDEḲ, which GOD dictated to Him; what He wrote upon is not said. None of the archangels or angels could hear what was said or see what was written, and the celestial hosts only heard this later from MARY. When JESUS gave the book to MARY He gave her directions as to its use, and explained to her its power thus: Its contents were to be revealed to believers only. The man who possesseth the book shall never go down into judgment, or into Sheol. The sins of the man who ties it to his neck, or carries it, shall be remitted, and the mere recital of the words in it during the Sacrament shall cause the sins of a man to be forgiven to him. And if they (*i.e.* the priests or relatives of the dead) make the Sign of SOLOMON'S SEAL thrice over the bier of a dead man on the day of burial, angels shall conduct him through the Gates of Life, and lead him into the presence of GOD in the kingdom of heaven. Thus the LEFÂFA ṢEDEḲ made a man pure and holy upon earth, and secured heaven for his soul.

The words of JESUS to MARY when He gave her the book have many parallels in the Rubrics to the Egyptian BOOK OF THE DEAD. If Chapter CI be written upon a strip of linen and laid upon the neck of the deceased on the day of the funeral, he shall join the followers of HORUS, and shall become a star in heaven, face to face with SEPṬIT (SOTHIS). A copy of Chapter CLVI attached to the neck of the deceased would open the gates of the Underworld to him. Chapter CLXI was a "great mystery," and was not to be revealed to the ignorant or those "who were outside," *i.e.* those who were not followers of OSIRIS; the recital of it enabled the soul to pass through the four entrances into heaven, but no man of another religion was to hear it recited. The recital of Chapter CLXIII made the soul immune from the attacks of SET (*i.e.* the Devil), and enabled it to enter into the most secret council-chamber of the god. And the knowledge of Chapter LXIV made a man to flourish in this world and in the next.

The allusion to the SEAL OF SOLOMON is unusual and interesting. Among the HEBREWS, SYRIANS, ARABS and ETHIOPIANS SOLOMON was always regarded as a mighty sorcerer, and among these peoples his reputation as a magician has always been greater than his fame as the wise and powerful king of ISRAEL. A widespread tradition says that he was the master of the winds of heaven, and could control their action, and that all birds, beasts, reptiles and fish were subject unto him. He was the overlord of ASHMEDAI,[1] the king of the devils and fiends, and he bound him fast with a chain on which the great Name of GOD YHWH was cut. Some say that the chain was in reality a net in which he caught the devils as a fisherman catches fish in a net. He imposed his will on every creature by means of a magical ring, on the metal or bezel of which the great Name of GOD was also inscribed, and he owed not only his position as king of ISRAEL, but also his very existence, to the possession of this ring. Some say that the Name YHWH only was cut upon this ring, but others say that the Name was placed within two interlaced triangles, which were arranged like the two triangles inside which the magicians of the Middle Ages wrote the magical word ABRACADABRA.

According to the Abyssinian legend of the QUEEN OF SHEBA SOLOMON gave his ring to the Queen just before she set out on her return journey, and she sent it back to him by the hand of her son MENYELEK when he made himself known to SOLOMON in JERUSALEM. Some modern ABYSSINIANS maintain that the design cut on the ring, or on its bezel, was copied by the ancient magicians of their country before the ring was taken back by MENYELEK,

[1] In a curious picture found in a manuscript in the possession of Dr. HERMANN GOLLANCZ, we see King SOLOMON mounted on a horse and in the act of spearing ASHMEDAI (ASMODEUS), who lies prostrate on the ground. For a facsimile see GOLLANCZ, *The Book of Protection*, Oxford, 1912, Plate facing p. 26.

and they assert that some of the curious figures inscribed on the parchment amulets which are found all over ABYSSINIA are copies of it. Nearly all the legends of the magical powers of SOLOMON are based upon various Tracts in the Talmûdh and other Jewish writings, and copious extracts from these will be found in EISENMENGER'S *Entdecktes Judenthum*, Theil I. pp. 351, 356, 357, 440 ff. For the story of how SOLOMON made a devil conduct HIRAM, king of TYRE, through the seven divisions of hell and bring him out from them safe and sound, see '*Emeḳ Hammelek*, fol. 112 (EISENMENGER, *op. cit.*, Theil II, p. 445).

The next sentence in the copy of the LEFÂFA ṢEDEḲ which we are describing shows that the book was not written expressly for the benefit of the devout man called STEPHEN, who says, "O bring thou me, thy servant STEPHEN, into the light of life, and into the salvation which is everlasting," for the name STEPHEN does not fill the blank spaces left for a name on Fol. 4*b*, col. 2, and Fol. 5*a*, col. 2. And on Fol. 6*a*, col. 1, where a blank space had been left, we find the name of "WALDA MÎKÂ'ÊL," which seems to suggest that two men, the one called STEPHEN and the other WALDA MIKÂ'ÊL, purchased the manuscript from a scribe who made a business of writing for sale copies of the LEFÂFA ṢEDEḲ, to which the names of the deceased might be added when their relatives purchased them. The same custom was common among the scribes of ancient EGYPT, who wrote copies of the BOOK OF THE DEAD, leaving in every chapter a blank space in which could be inscribed the name of the men or women on whose behalf they were purchased. The EGYPTIANS, like the ABYSSINIANS, also interpolated prayers in the texts of the chapters, and the substance of them closely resembles that of the Christian prayers. Thus in Chapter CXXVI we have, " Wipe out my evil deeds, and put away my sin, and let there be nothing on my part to prevent this. Grant that I may traverse the Ammaḥet and Re-stau, and pass through the hidden gates of Amentt (*i.e.* the kingdom of OSIRIS). As food and drink are given to the living Spirits, so let them be given to me."

When JESUS had finished this description of the power of the LEFÂFA ṢEDEḲ He revealed to Mary a series of nineteen names, the utterance of which would secure life and salvation for a man, whether living or dead; and He told MARY that men were to " take refuge," or put their trust in them. Another prayer by STEPHEN follows, and after that come a series of fifty names. A few of the names are derived from Hebrew, *e.g.* 'AMÂNÛ'ÊL = EMMANUEL, and it is probable names ending in *êl* or *îl*, *e.g.* 'ADNÂ'ÊL and BADMÂHÎL, are corruptions of the names of Hebrew angels or archangels. " 'Alfâ " is probably " Alpha," and " 'A'ô " seems to represent " Alpha + Omega." The other names may represent the various powers or attributes of our Lord, but it is more probable that many of them are garbled forms of the names of the emanations, and Aeons, and angels which we find in such works as the *Pistis Sophia*, and the Gnostic work the *Book of Iêu* and such-like. These made their way into ABYSSINIA in the magical writings of the COPTS, who in their turn derived them from Greek or even oriental sources. Some of the names may be of native origin, and the inventions of Abyssinian magicians. It may be noted that the ancient Egyptian theologians declared that "the gods" were merely personifications of the various names of the "ONLY ONE" god, whether he was called TEMU, or KHEPERA, or RĀ, or ÅMEN. And RĀ had seven souls, and fourteen "Kau" (*i.e.* doubles) that were called Ḥeḳ, Neḳht, Aakhu, User, Uatch, Djefa, Sheps, Senem, Sepṭ, Ṭeṭ, Maa, Setem, Sáa, Ḥu.

When Jesus had revealed these names to MARY He told her that the mercy of GOD was full and perfect, and that if men believed in His Name the judgment which He would pass upon them would give them life and salvation. With two prayers, one by STEPHEN and one by WALDA MÎKÂ'ÊL, the FIRST SECTION comes to an end.

THE SECOND SECTION

The SECOND SECTION opens with the words, "In the Name of the Father and the Son and the Holy Ghost, One God," and then goes on to say that "JESUS wrote with His own hands the following names." Among the forty-four names which are then given are the names of the letters of the Alphabet, from 'Alîf to Tâw! It is interesting to note that the names of the letters are given in the order of the letters of the Hebrew, and not the Abyssinian, alphabet. A short prayer follows by STEPHEN, who entreats GOD that his petition may reach Him, and that he may never see the smoke of the fires of hell.

The next paragraph is an address to the great and everlasting GOD, but by whom is not stated, and contains a petition that the magical names which our Lord revealed to the divine PETER may be revealed [to him or her].

This petition is followed by sixty-nine magical names, some of which, *e.g.* YÂW, 'ÊLÔHÊ and 'AMÂNÛ'ÊL, are derived from Hebrew, and the others from sources unknown to me. It is quite clear that they were not invented by any modern Abyssinian, and that they were borrowed by the compiler of the LEFÂFA ṢEDEḲ from some early Christian (Coptic) magical work. It is possible that they were taken from some portion of a work like the *Apocalypse of Peter*, which the reader will find discussed and fully described by M. R. JAMES in his *Apocryphal New Testament*, Oxford, 1924, p. 505 ff. Or they may have been taken from some copy of the 'ARDE'ĔT or magical prayers which CHRIST taught His disciples. The object of these prayers was to save the disciples from every evil and every kind of disease, from the poison of serpents, from enemies, from the spells of magicians, from the curses of sorcerers, from devils and phantasms of darkness, from death and the flames of hell, and from the Arch-devil DIABOLOS. After each of these a number of magical names are given, and among them are some which are found in the LEFÂFA ṢEDEḲ (see Brit. Mus., MS. Add. 16,245, fol. 7 ff., and Add. No. 24,996; DILLMANN, *Catal.*, p. 61; WRIGHT, *Catal.*, p. 112). Following the sixty-nine names in the LEFÂFA ṢEDEḲ is a declaration by WALDA MÎKÂ'ÊL that he takes refuge in these names.

After this comes the following spell, which is repeated in other places in the LEFÂFA ṢEDEḲ :

SÂDÔR	'ALÂDÔR	DÂNÂT	'ADÊRÂ	RÔDÂS
ሳዶር፡	አላዶር፡	ዳናት፡	አዴራ፡	ሮዳስ፡

These words are said by the Abyssinians to be the names of the five nails which were driven into our

Lord when hanging on the Cross, but LUDOLF pointed out (lib. III. chap. 4, No. XXXV. p. 351) that they were merely a faulty transcription of the old, well-known palindrome

SATOR AREPO TENET OPERA ROTAS

Ancient sorcerers attached great importance to magical formulæ which read the same from either end, and this is a classical example of such formulæ. Palindromes are said to have been invented by SOTADES, a native of MARONEIA in THRACE, who flourished in the first half of the third century B.C. He attacked PTOLEMY PHILADELPHUS in certain obscene poems on the occasion of the king's marriage to his sister ARSINOË, and was cast into prison. Later he escaped from ALEXANDRIA, but was caught by PATROCLUS, one of PTOLEMY'S generals, who shut him up in a leaden coffin and cast him into the sea. The above palindrome has been found in many places on the Continent, and TREVELYAN, in his *Folklore of Wales*, p. 233, states that a copy of it, cut upon a stele of the Roman Period, was found in Glamorgan in 1850. It was arranged in the form of a magical square thus,

S	A	T	O	R
A	R	E	P	O
T	E	N	E	T
O	P	E	R	A
R	O	T	A	S

and its recital was supposed to cure the bite of a mad

dog. The five words, which were said to represent the five wounds of CHRIST, were to be written on a crust of bread, and this was to be applied three times to the wound caused by the dog. Also the Lord's Prayer was to be recited five times, once for each of the five wounds of our Lord (see ELWORTHY, *The Evil Eye*, London, 1895, p. 401).

This palindrome passed into EGYPT, probably in some magical work written in Greek, and was adopted by the COPTS, perhaps in the sixth century (BASSET, *Apocryphes*, Pt. V, p. 5), but KRALL would not admit that its adoption took place earlier than the eighth century (RAINER, *Mittheilungen*, Bd. V.). Its form in Ethiopic, as given above, shows that it came into ABYSSINIA through the Coptic from EGYPT, but whether it entered the country by way of the NILE and NUBIA, or whether it was brought in by the PORTUGUESE, as WORRELL thinks (*Zeit. für Assyr.*, Bd. XXIX. p. 89), is uncertain. The Ethiopic version of the PRAYER OF MARY in BARTOS contains the palindrome, but it is wanting in the Coptic version published by CRUM from the Brit. Mus. MS. Oriental 4714 (see " A Coptic *Palimpsest* " in *Proceedings Soc. Bib. Arch.*, Vol. 19 (1897), p. 210). For the Ethiopic text of the Virgin's Prayers see CONTI ROSSINI, *Acad. dei Lincei*, Rendiconti, Series V., Vol. V. p. 455 ff., and for a French translation see BASSET, *Apocryphes*, Paris, 1895, p. 11 ff. An English translation from MSS. in the British Museum is given on pp. 95 and 112 ff.

The palindrome " Sator Arepo Tenet Opera Rotas " is to me meaningless, but in its complete form it is, according to HEIM, R., " Incantamenta Magica " (in the *Jahrbücher für Class. Phil.*, Leipzig, 1893, p. 463 ff.), the remains of a solemn hymn which the early

ROMANS used in their religious exercises. It is to be completed thus :—

SAT ORARE POTENter
ET OPERAre RatiO TuA sit.

See also SCHWARTZ, "Der Zauber des 'ruckwärts' Singens und Sprechens" (in *Indogermanischer Volksglaube*, p. 257).[1]

The SECOND SECTION concludes with the words, "I, thy servant STEPHEN, take refuge in the five nails of the CROSS of our Lord JESUS CHRIST." Below them is an elaborately decorated figure of a cross with two horizontal bars, but whether this is intended to belong to the SECOND SECTION or to the THIRD is not clear.

THE THIRD SECTION.

The THIRD SECTION begins with the usual, "In the Name of the Father and the Son and the Holy Ghost, One God."

[1] Palindromes in English are not common, and the two most commonly quoted are :—

1. MADAM I'M ADAM.
2. LEWD DID I LIVE & EVIL I DID DWEL.

In French we have :

L'AME DES UNS IAMAIS N'USE DE MAL.

In Latin :

1. ROMA TIBI SUBITO MOTIBUS IBIT AMOR.
2. SI BENE TE TUA LAUS TAXAT SUA LAUTE TENEBIS.
3. ARCA SERENUM ME GERE REGEM MUNERE SACRA.
4. SOLEM ARCAS ANIMOS, OMINA SACRA, MELOS.
5. ACIDE ME MALO, SED NON DESOLA ME MEDICA.
6. ABLATA AT ALBA.
7. SI NUMMI IMMUNIS (A LAWYER's motto, "Give me my fee, I warrant you free."

In Greek : Νίψον ἀνομήυα μὴ μόναν ὄψιν

(WHEATLEY, H. B., *On Anagrams*, London, 1862.)

The first paragraph contains a prayer, *i.e.* spell, which is to be recited when the deceased is being borne to the tomb. It contains six magical words or names, and reads DEḲÂS BATRÔN KÛGÛYÂ GÂNÔN KÂWES ḲÎREL. It is followed by a statement, which may be described as a Rubric, and which declared that the deceased for whom the prayer shall be recited on the last day, *i.e.* the day of the funeral, shall not be attacked by anything [evil or harmful]. With this Rubric may be compared the Rubrics of some of the chapters of the BOOK OF THE DEAD. Thus we have: "This chapter shall be recited over a Ṭeṭ of gold. . . . And it shall be placed at the neck of the deceased on the day of the funeral. If this amulet be placed at his neck he shall become a perfect (or honourable) spirit in the Underworld" (Chapter CLV). Compare also the Rubrics to the four following chapters.

The next paragraph mentions GOG and MAGOG, and speaks of the coming of the "son of SATAN," *i.e.* ANTICHRIST. GOG and MAGOG, according to METHODIUS, Bishop of PATARA in the fourth century, and their kindred peoples, were descendants of JAPHET, and lived on the confines of the East. Their appearance was hideous, and they were more wicked and unclean than any other dwellers in the world. They were as ignorant as the beasts, they knew not GOD, and they lacked the power of reason; they ate mice, snakes, scorpions and every kind of reptile, and they did not bury the bodies of their dead, but ate them. ALEXANDER THE GREAT, seeing their wickedness, prayed to GOD, and then built a gate of brass at the entrance of a defile which was formed by the two mountains which GOD had made to approach within twelve cubits of each other, and so shut in these filthy peoples. The words which ALEXANDER used against them are quoted in a parchment amulet described by WORRELL (*Zeit. für Assyriol.*, Bd. XXIV., p. 78). The names of the peoples who were imprisoned within this northern gate are preserved by SOLOMON, Bishop of AL-BAṢRAH, in his *Book of the Bee* (ed. BUDGE, p. 128), and are as follows: Gôg, Mâgôg, Nâwâl, Eshkenâz (Ashkenaz), Denâphâr, Paḳṭâyê, Welôṭâyê, Humnâyê (Huns), Parzâyê, Daḳlâyê, Thaubelâyê, Darmeṭâyê, Kawkebâyê, Dogmen, Emderâthâ, Garmîdô, Cannibals, Therḳâyê (Thracians), Ālânâyê (Alani), Pîsîlôn, Denḳâyê, and Salṭrâyê. At the end of the world, when all peoples are at peace, these nations shall force their way through the gate of brass, and lay waste the earth. They will eat men, women, children, cats, dogs, and reptiles, and having laid waste and ravaged the whole earth for one week, they will all gather themselves together in the plain of JOPPA, and then the hosts of the angels will descend from heaven and destroy them (see BRANT's edition of METHODIUS, p. 20). A week and a half after the destruction of those filthy peoples, the son of perdition, *i.e.* ANTICHRIST, shall appear. As soon as he is revealed the king of the Greeks will go up and stand on GOLGOTHA, and set the royal crown upon the top of the HOLY CROSS, on which our Lord was crucified; the CROSS and the crown will be taken up into heaven, and the king will die forthwith. This king will be descended from KÛSHATH, the daughter of PÎL, the king of the ETHIOPIANS; for ARMELAUS (ROMULUS), the king of the GREEKS, took KÛSHETH to wife, and the seed of the ETHIOPIANS was mingled with that of the GREEKS.

From this seed a king shall arise who shall deliver the kingdom over to GOD, as the blessed DAVID has said, "CUSH will deliver the power to GOD" (Psalm lxviii. 31). When the Cross is raised up into heaven, every king and governor will be brought to nought, and GOD will withdraw His providential care from the earth. Then shall the "son of SATAN" appear.

SATAN wished to follow the example of the Almighty, and to send a son into the world to combat righteousness, and to pretend to be CHRIST. He was unable to find a virgin for his purpose, and he begot his son by a married woman of the tribe of DAN; this son was conceived in CHORAZIN; born in BETHSAIDA, and reared in CAPERNAUM, and for this reason our Lord proclaimed: Woe to these three [cities] in the Gospel (Matt. xi. 21). This "son of SATAN" shall lead astray the world, for he shall show deluding phantasms of miracles, the blind seeing, the lame walking, the lepers cleansed, the sun becoming black, the moon changing its appearance, etc.; but he shall not be able to raise the dead. He will sit on a throne in the Temple at JERUSALEM and will say, "I am the Christ, I am God, I am the fulfilment of the types and parables." He will be borne aloft by legions of devils like a king and a law-giver. He will be made a dwelling-place for devils, and all Satanic workings will be perfected in him. And when every one is standing in despair then will ELIJAH come from Paradise and convict the deceiver.

Now it is quite clear that the author of the LEFÂFA ṢEDEḲ was acquainted with the legend of GOG and MAGOG, and the prophecy about the coming of the "son of SATAN," as set forth by METHODIUS, but whether he derived his information from a Coptic or an Arab source cannot be said. In the paragraph following the mention of ELIAS, the doom of the man who believes in the "son of SATAN" is clearly foretold, and is sharply contrasted with that of the believer in CHRIST, who shall not only escape from punishment, but shall be held worthy to walk with the HOLY GHOST.

Taking the text of the LEFÂFA ṢEDEḲ as it stands, it is difficult to make the next paragraph fit the context. After the words "God saith, I am the God of the heavens and the earth," come the words, "And NÂTNÂ'ÊL the King shall go about himself." Next we have the words, "The Christian shall lack (?) (or lament?) the tunic (*kalamîdâ*), the fountain (or spring) of glory and life. This is he who shall ride the horses of life." But who is NÂTNÂ'ÊL (NATHANIEL) the King? Is it possible that NÂTNÂ'ÊL is a scribe's mistake for SÂṬNÂ'ÊL, *i.e.* the Devil? The allusion to the Christian is not clear, though the meaning of each word is, and it seems doubtful who is to ride the "horses of life on the day of reward and judgment" (or punishment). The final words of the paragraph show that the day of judgment is referred to, for they read, "And in that day the sun shall become black, and the moon shall become blood" (see Joel. ii. 10, 31; 3, 15; Matt. xxiv. 29; Acts ii. 20; Rev. vi. 12).

The next two paragraphs are prayers by WELDA MÎKÂ'ÊL and STEPHEN. These are followed by a conversation between GOD and MICHAEL, the "Angel of God," or the "Angel of the Face," that is to say, the greatest of all the angels. A sound as of thunder reaches MICHAEL, and he asks GOD what it means? And in answer GOD tells him that the noise comes from the place where the souls of sinners and those who treated His word with contempt are suffering punishment. And GOD assures MICHAEL that the man who has a copy of the LEFÂFA ṢEDEḲ written for him and wears it round his neck is blessed, *i.e.* shall be immune from the punishment of GEHENNA. And in that day there shall be a sun that shall not set, and a lamp that shall not be extinguished, and the sound of the reward [of the blessed] that shall never cease; And the kingdom of GOD that shall never be destroyed, and His fourfold (?) fire-crowned throne that shall never be overthrown.

Then the Angels ask GOD to declare to them His name, so that they may praise and hymn Him. And GOD gives them His Seven Great Names, that is to say, the names of His Seven principal Characters or Aspects, viz.,

'Îyâwâdâ.	Kînyâ.	'Amânû'êl.
'Îyâsûs.	Kerestôs.	'Îyâd.
	'Ĕgzî'abeḥêr.	

The man who puts his confidence in these seven names (which may be compared with the Seven Souls of the Egyptian Sun-god RĀ), shall escape from the devouring everlasting fire, and the Worm that never sleeps. In a further address to MICHAEL GOD again declares the efficacy of the LEFÂFA ṢEDEḲ in procuring for the man who possesses the book immunity from hell fire. The "water of his prayer" probably refers to the consecrated mixture of oil and water, *i.e.* holy water, which was used in connection with the recital of prayers and magical spells generally. Its composition is attributed to CYRIL, Archbishop of JERUSALEM (CRUM, *Proc. Soc. Bibl. Arch.*, Vol. XIX. p. 211). The ancient EGYPTIANS made use of holy water in their rituals, but the cleansing and sanctifying element in it was natron.

The Worm that never sleeps finds its prototype in ancient Egyptian texts. The BOOK OF THE DEAD (Chap. I B) says that there were Nine Worms that lived in the Ṭuat, and devoured the souls and bones and blood and bodies of all the men and women who came there, both living and dead. Their names were: (1) Narti-ānkhi-em-senu-f, (2) Her-f-em-qeb-f; (3) Ānkhi-em-fenṭu; (4) Sām-en-qesu; (5) Ha-huti-ȧm-sau; (6) Shep-timesu; (7) Ȧmi(unemi?)-sāḥu; (8) Sām-em-snef; (9) Ānkhi-em-betu-mitu. But of all Worms the most terrible was he who dwelt in the bight of the River of Hell, and passed all his days and nights in devouring the souls of the dead; he never slept, and his jaws never ceased from their horrible work.

The oil which was mixed with the "prayer water" was, when obtainable, the famous Mêrôm oil. It was made from the balsam plants which grew round about the Well of the Sun ('Ain ash-Shams) at HELIOPOLIS. Tradition says that the Virgin MARY threw the water from the bath in which she had washed our Lord out on the ground near their tent, and that balsam-bearing plants immediately sprung up there. In all magic and religious ceremonies oil played a prominent part. During the performance of the ceremony of "Opening the Mouth" the EGYPTIANS anointed the statue of the deceased in the Ṭuat Chamber with the SEVEN HOLY OILS, the names of which were: (1) Seth-ḥab; (2) Ḥeknu; (3) Sefth; (4) Nem; (5) Ṭuaut; (6) Ḥa-āsh; (7) Ḥatt-ent-Theḥnu.

When MICHAEL had thanked GOD for describing to him the things that shall take place at the last day, he and all his angels gathered themselves together in order to hear CHRIST read to them the contents of the book LEFÂFA ṢEDEḲ, which had been dictated to Him by His Father. The book was sealed with the triple Seal of the TRINITY, and the only beings who were authorized to break the seal and read therein were the TWENTY-FOUR PRIESTS of heaven,[1] and the FOUR EVANGELISTS. The Priests, according to Rev. iv. 4. 10, were clothed in white raiment, and had crowns of gold upon their heads. The Four Evangelists took the book, and broke the triple seal, and having looked therein they read out aloud its contents, so that all the angels might hear. Then seven angels took trumpets and blew blasts on them, and seven other angels took vessels [of water?] and poured them out on the face of the earth, for the sanctification of the good and righteous men that were thereon. Through this the souls of the righteous men became free to traverse the heavens and the earth, to pass through the Seven Gates, and the Seven Light Spaces, and ceased to be under the authority of the Seven Bearers of the Throne of GOD. In this way was the Awful Name of GOD made known to the Prophets and the Apostles, each in the place where he was. These things took place probably on the sixteenth day of the month MASKARAM (Sept.), the day on which, according to the SYNAXARIUM,[2] the festival of the discovery of the CROSS by Queen HELENA, and the consecration of the Temple and Church of the Tomb of CHRIST, were celebrated. After the emptying of

[1] Their names were: 'Akîyâl, Fânu'êl, Ḳartîyâl, Dartîyâl, 'Îlyâl, Zartîyâl, Tîtâ'al, Yûyâl, Kartîyâl, Lebtîyâl, Mîtâ'al, Mîrâ'al, 'Aûktîyâl, Bîtâ'al, Râûâl, Sarwâl, Sakarwâl, 'Aksîfâ'al, 'Anîwâl, Fîlalê'al, 'Akerstîyâl, 'Aksîfâ'al (sic), 'Aûnûâl. [One name is given twice, and two names are wanting.]

[2] *I.e.* the *Maṣḥafa Senkèsâr* of the ETHIOPIANS.

the seven vessels on the earth GOD declared to His saints the twenty names which formed the component parts of His name.

After a short prayer in which WALDA MÎKÂ'ÊL prays that GOD will make him to ascend into heaven, even as He made MARY to ascend into heaven, we find seven other names of GOD, which are said to be unknown to men, and to have come forth from the mouths of the Father and the Son and the Holy Ghost, in their own speech. These are followed by twenty more names which are said to be the "keepers of the soul and the gates thereof." He who carries them on his person silently, and with patient humility, and repeateth them in a humble voice, and in the fear of GOD, shall be saved. And he who lends a ready ear to these words shall prosper in this world, and he shall traffick in gold, and silver, and costly stuffs, but the man who turns a deaf ear to them shall become a slave of DELESḲEYÂM (?). And the man who knows the name which JOHN bestowed upon CHRIST when he baptized Him shall neither see hell, nor suffer in the place of torment, and GOD will show mercy upon him. Then follows a prayer in which STEPHEN prays that he may be made to ascend into heaven as MARY was made to ascend there, and after this come three groups of magical names, containing sixteen, seven and ten names respectively. STEPHEN says that he takes refuge in these in order to prevent death and suffering coming upon him. And WALDA MÎKÂ'ÊL beseeches our Lord by these names and by the blood of GEORGE [the martyr] to remember him when He comes into His kingdom.

This SECTION ends with a repetition of the names of the five nails of the CROSS, and a prayer to our Lord for everlasting remembrance. On Fol. 13*a* are drawn two crosses with elaborate decorations.

The Fourth Section

This Section begins with the usual " In the Name of the Father," etc. The first paragraph mentions that God gave the Lefâfa Ṣedeḳ to Mary as a covenant for the last day, *i.e.* the Day of Judgment, and that she carried the above magical names in her womb as a protective covering (literally " helmet "). The man who carries these names within him, like Mary, or ties a fillet inscribed with them to his person, shall never see Gehenna, and shall find life everlasting.

The next few paragraphs describe a dialogue which took place between Jesus and Mary. Our Lord had communicated to her the words of the Lefâfa Ṣedeḳ, and all the mighty names of God which He had revealed to Him, and to Michael and his angels, but Mary was not satisfied that these would procure the escape of the members of her family from the Judgment and from the devouring fire of Gehenna. Therefore she asked our Lord to tell her which was the greatest of all His names, and He promised to reveal to her the names which were " difficult for the hearing, and were hidden from the sight," and which would keep in safety the man who was able to hear them. But apparently He did not do so. Mary then repeated her request, and besought Him to tell her the hidden or secret name of God. Again Jesus promised to tell her His name " correctly," but warned her that it was not to be regarded lightly, adding that the name was a difficult one for the unbeliever, and that it was unseemly to reveal it to the man who could not hear it. On this MARY promised not to reveal the names to foolish men, or to men of no understanding, or to those who did not wish for heaven, or to those who had not withdrawn themselves from earthly honours. In reply JESUS told her that He wished men to know the names which He would reveal to her, and then, standing on a pillar of cloud, and enveloping Himself in a flame of fire, He revealed to her Three Three-fold Names, which were, presumably, the secret or hidden names of the Father, and the Son, and the Holy Ghost. These are followed by petitions to ten Archangels, whose names are given by STEPHEN and WALDA MÎKÂ'ÊL, who declare that they take refuge in the names of the Four Beasts, the Throne and City of God, MARY, the Evangelists, the Prophets, the Apostles, the Priests and Soldiers of heaven, the Seventy-two Disciples, the Three hundred and Eighteen Fathers of the Council of NICEA, and the angels of heaven. This SECTION ends with the statement that STEPHEN has taken refuge in the names of the five nails of the CROSS OF CHRIST, and the names of which, Sâdôr, 'Alâdôr, etc., are repeated.

An interesting parallel to the persistent request of MARY to JESUS to reveal to her His secret name is found in the ancient Egyptian *Legend of* RÂ *and* ISIS. The parallel is important, too, for it shows that both ISIS and MARY believed that their God possessed a secret name, by the use of which He created the world and governed it. According to the Egyptian legend, RÂ, the self-begotten and self-created god, the creator of heaven and earth, and of every being and thing in them, possessed "many names," which were unknown even to the gods. The goddess ISIS saw RÂ exercising his powers daily, and she wondered if it were possible to become like unto that god, and to make herself mistress of heaven and earth. She pondered deeply on the matter, and decided that she could make herself equal to the god if she could only gain possession of the secret name of the holy god. As the god was passing across the sky some of his spittle fell on the ground, and Isis took it up and mixed earth with it and fashioned a serpent, on which she, being a great magician, bestowed magical powers. And she placed this serpent on the path of the Sun-god and departed. On the following day the god passed over the path by which the serpent lay, and as he did so the reptile bit him, and straightway the heat of life began to diminish in the god's body. As the venom flowed through his body his members quaked, his jaw-bones rattled together, and he began to suffer excruciating pains. He cried out to the gods whom he had created saying, "I am a king, the son of a king, the essence produced by a god. I am the Great One, the son of the Great One. My father devised for me my name. I am of many names, and many forms, and my substance existeth in every god. My name was bestowed upon me by Tem and Horus, the gods who devise and assign names. My father and my mother pronounced my name, and he who begot me hid it in my body (or belly) so that he who wished to work magic upon me by means of his magic would not be permitted to gain any power over me." At the cry of Râ all the gods crowded about him and began to weep, but meanwhile the poison was carrying out its deadly work in the body of Râ, and his collapse became imminent.

Then came Isis, who was the mistress of spells, the utterance of which would drive away every disease and restore the dead to life, and having told RÂ exactly what had happened to him, she said, "This attack can be overthrown by means of beneficent magic; I myself will remove the calamity from thy sight." RÂ was proceeding to describe his sufferings, when ISIS interrupted him and said, "Tell me thy name, O divine father, for a person maintaineth his life by means of his name." In answer RÂ continued to enumerate his titles, and to describe his powers at length, but meanwhile, as that text pithily remarks, "The progress of the poison in the god's members was not checked, and his pains were not relieved. Again ISIS spoke, and she said to RÂ, "Thy name is not among the words which thou hast uttered. Tell me thy name, and the poison shall depart, for whosoever shall declare his name shall live." Whilst she was saying these words, the poison inflamed the body of the god more and more, and the burning pain it caused was worse than the burns caused by fire. At this moment RÂ surrendered, and permitted ISIS to search through his body and to transfer his name from his own body to that of ISIS, and he withdrew himself from the sky so that the gods might not know what was taking place between ISIS and himself.

It will be remembered that GOD hid himself behind seven curtains of fire when He was dictating the LEFÂFA ṢEDEḲ to JESUS, and shrouded Himself in a cloud of light, and that CHRIST enveloped Himself in a flame of fire when He revealed His secret name to MARY. When the secret name of RÂ had been taken from his body by ISIS, the great lady of magic uttered the following spell: "Flow out, poison, eject thyself from RÂ. Come forth, Eye of Horns, who proceeded from the god, fashion firmly for him (*i.e.* RÂ) his mouth. I work, I come to make the poison to fall down on the earth, for it hath been overcome. Indeed the name of the great god hath been lifted from him. RÂ liveth, the poison dieth; the poison dieth, RÂ liveth." Thus ISIS used the secret name of RÂ as a magical spell, and made him to recover from the bite of the snake; in the same way MARY and the Apostles used the secret names of the PERSONS of the TRINITY to heal the sick and to raise the dead.

[For the text and translations see PLEYTE and ROSSI, *Papyrus de Turin*, foll. 31, 77, 131–138; LEFÉBURE, *Zeitschrift Aeg. Sprache*, 1883, p. 27; BUDGE, *First Steps in Egyptian*, pp. 241–256.]

The Virgin MARY plays in the LEFÂFA ṢEDEḲ the part which ISIS plays in the BOOK OF THE DEAD. From first to last ISIS was regarded by the EGYPTIANS as a friend of the dead. She was a mistress of magic, *ḥeka*, and she always employed her great power in helping both the living and the dead. By the spells which she knew how to utter fluently and correctly, and with the proper intonation, she gave her dead husband OSIRIS power to beget his son HORUS. She restored HORUS to life after he had been stung to death by the scorpion sent to him by SET, the arch-god of evil; and she assisted the blessed dead in their efforts to enter the kingdom of OSIRIS, and fed them with celestial food daily in the presence of OSIRIS. The spells which she used she had learned from THOTH, the heart or tongue of the Great God, or from the Great God himself, even as MARY learned the magical names of GOD Almighty from our Lord. The legend of OSIRIS says that after his murder by SET he was obliged to submit to the ordeal of judgment by the great gods of heaven, but ISIS was not tried in the

HALL OF JUDGMENT, and when OSIRIS became king and god of the dead, she took her stand, together with her shadowy counterpart NEPHTHYS, by the side of OSIRIS as he sat on the throne of judgment in the Hall of Maâti without any opposition on the part of the gods. MARY likewise escaped the Judgment, and was taken up to heaven and was seated side by side with the Father on the Throne of Heaven.

THE FIFTH SECTION

The first paragraph of this Section is a prayer or spell, the recital of which would, it was believed, enabled a man's soul to pass through the earth and travel without hindrance or obstruction to heaven. The dynastic EGYPTIANS, COPTS and, it seems, ABYSSINIANS, all believed that the soul on leaving the body set out on a long and difficult journey through the earth in order to reach heaven. The pagan EGYPTIAN sought the heaven of OSIRIS, and the Christian EGYPTIAN and the ABYSSINIAN the heaven of CHRIST. Everywhere on the road the angels of darkness and devils lay in wait to pounce upon the soul in order to obstruct its passage or to kill it. The Egyptian protected himself with the spells found in the BOOK OF THE DEAD, and appealed to his gods for protection. In the Rubric of Chapter CLVIII we read: " This Chapter shall be written upon a bandage of stout linen which is to be wrapped about every limb of his body. Then the deceased shall not be turned back at any gate of the Ṭuat; he shall eat, and drink, and ease himself even as he did when he was upon earth; none shall rise up to cry out against him, and he shall be protected from the hands of every enemy for ever and ever. If this writing be recited on his behalf on earth, he shall not be seized upon by those sent to attack him in all the earth. Wounds shall not be inflicted upon him, he shall not be slaughtered by SET, he shall not be carried away into captivity, but he shall enter the Court [of OSIRIS] in triumph." The ABYSSINIAN believed that this spell, written on a strip of linen, whether attached to his body after death, or recited on his behalf after his burial, would do for him exactly what the spell in the BOOK OF THE DEAD did for the EGYPTIAN. For the name of OSIRIS he substituted that of CHRIST, and MICHAEL, GABRIEL and the PARACLETE take the place of the gods of the Seven Ārits and the Pylons. The title of LAMP applied to CHRIST is of interest, for in all Egyptian and Nubian magical ceremonies the lighted lamp played a prominent part, and the magician stood with a lighted lamp on his right hand and a censer filled with burning incense on his left.

Prayers by STEPHEN follow, and then we are told that GOD spake unto the Twelve Apostles and to the Seventy-two Disciples, and commanded them to make copies of the LEFÂFA ṢEDEḲ, and to recite the work to every Christian they met. The possession of a copy of the book carried with it immunity from the terrors and punishments of GEHENNA. STEPHEN then points out that, as there is no tree the wood of which when burnt will not produce smoke, so there is no man who hath not committed sin. And WALDA MÎKÂ'ÊL is consoled by the fact that it is the same book that drives away devils from GOD's Throne, and from his own soul, viz. the LEFÂFA ṢEDEḲ. Next we have the SEVEN MAGICAL NAMES OF CHRIST, among them being His baptismal name, and the name of a personification of His strength, and the name by the utterance of which He broke down the gates of hell, and smashed their bolts; the last three names are unexplained. The section ends with a repetition of the names of the five nails of the Cross of CHRIST.

THE SIXTH SECTION

This section opens with the statement that the Disciples urged our Lord to reveal to them His secret or hidden name, for they wished to know the name by virtue of which He existed and came into being. At length He answered them, and after commanding them to guard and preserve the Book of LEFÂFA ṢEDEḲ, and describing to them the benefits which would accrue to the man who had a copy made of it, He revealed to them His secret or hidden name, which was known only to the Four and Twenty priests of heaven and to MARY, the Virgin. Then follow forty-two names and three triple names, which He said were the greatest of all His names. He then declared to the Disciples that it was by this name alone that they and mankind in general could be saved. The rest of the Section consists of a long speech by our Lord in which He describes the powers of the Book of LEFÂFA ṢEDEḲ.

THE SEVENTH SECTION

In this section there is an allusion to the old legend in which our Lord is said to have dispatched St. ANDREW, the Apostle, to the CITY OF THE CANNIBALS, where MATTHIAS was imprisoned, and commanded him to release him. ANDREW in reply pointed out that it would take him two years to travel to the city, and that a great sea flowed between that city and the place where he was. When ANDREW was ready to set out on his way our Lord revealed to him the Six triple names which GOD the Father used before He made the heavens and the earth, and the Eight triple names which belonged to Himself, and the Seven names of the Holy Spirit, and said to him, "Pray ye in these my names, and the gates shall be opened and those who are therein shall be set free." The legend, which is printed on p. 91 ff. goes on to say that ANDREW obeyed the Lord's command, and that he broke into the prison in the CANNIBAL CITY and set free MATTHIAS, by means of the use of these magical names of Christ. In the last part of the section is a group of Eight magical names which will protect a man from the EVIL EYE, and from SATAN and his devils, and CHRIST is entreated by WALDA MÎKÂ'ÊL to fetter his foes, even as he fettered the fiend BERYÂL in hell.

THE EIGHTH SECTION

In this section are given: (1) A series of thirty-four single magical names; (2) the names of ALPHA and ŌMEGA; (3) a series of Seven Sevenfold magical names; and (4) a series of One hundred and forty-one magical names. These are to be repeated by a man to guard him from the approach and attacks of the DEVIL.

THE BANDLET OF RIGHTEOUSNESS

Translation

THE BANDLET OF RIGHTEOUSNESS

[Fol. 2*a*] IN THE NAME OF THE FATHER, AND THE SON, AND THE HOLY GHOST, ONE GOD.

A prayer for salvation [from] MAṢḤAFA ḤEYWAT (*i.e.* the BOOK OF LIFE), which is called "LEFÂFA ṢEDEḲ," and which the Father wrote with His own hands before CHRIST was brought forth by the holy woman the Virgin MARY. It will make a man to enter the narrow gate, and make [him] to arrive in the kingdom of heaven, and guide him to righteousness (or, the truth). And this [book] is what CHRIST spake unto MARY, His mother, after He had been brought forth [Fol. 2*b*] by her.

THE FIRST SECTION

On the sixteenth day [1] of the month of Yakâtît (February 5–March) CHRIST appeared to MARY in the place where the righteous have their habitation in the Garden (*i.e.* Paradise), and in the place where sinners dwell in torment in hell. And when she saw it she was stupefied and trembled, and she feared with a great fear. And our Lady MARY spake [to Him]. And JESUS said unto MARY, "Fear thou not, O MARY, My mother, who didst carry me in thy womb, and didst bring me forth by the HOLY GHOST." And she said unto Him, "Wherefore did I carry Thee? Tell me, O my Son, how my kinsfolk are to be saved from this

[1] The Synaxarium says that on this day a festival in honour of the Virgin MARY is celebrated among all Christian peoples.

devouring fire? I am afraid for my own soul, and for [Fol. 3*a*] 'ÎYÂḲÎM (JOACHIM), my father, and for ḤANNÂ (ANNE or HANNAH), my mother, and for SÂMÛ'ÊL and YÔSÎF (JOSEPH), my brethren, and for 'ÊLSÂBÊT (ELISABETH), my sister, and for DÂWÎT (DAVID), the ancestor of my family. And now, tell me, O my Son, clearly and certainly, by what means these are to be saved from this devouring fire."

And JESUS said unto MARY, "I cannot declare [this] to thee, for the matter which is discussed by two [people] will go forth to a third person, and after him it will be sown broadcast among all men. And they will commit sin, saying, 'There are means whereby we may be saved.'"

And again MARY asked Him, and said unto Him, "Wherefore (or, to what end) did I carry Thee in my womb for nine months and five days?" [Fol. 3*b*.] And our Lady MARY wept bitter tears, and CHRIST wept with her.

And JESUS said unto her, "Weep thou not, O MARY, my mother, behold I will speak to my Father. And after He hath given me permission to do so I will tell thee."

And JESUS went to His Father, and He said unto Him, "Behold, MARY, my mother, is weeping. Give me the MAṢḤAFA ḤAYWAT (*i.e.* the BOOK OF LIFE), which Thou didst write with Thy holy hand before I myself was brought forth by MARY, the Virgin, [who now] sitteth upon her chariot of the Kîrûbêl (Cherubim), Thy throne."

And His Father said unto His Son, "Behold, I have given it unto Thee. Go Thou and say unto MARY, Thy mother, that I have hidden (or, will hide) from her nothing whatsoever; and so far as Thou art concerned it is fitting that I should reveal [Fol. 4*a*] unto Thee everything."

And JESUS wrote with a pen of gold. And a light cloud came and hovered over them, and they (*i.e.* GOD and CHRIST) made seven pavilions (or veils) of fire [round about them], and none knew and none heard, neither the angels nor the archangels, until they had told MARY the whole of the following words.

And [CHRIST] said unto her, "Take this [book] which I have given unto thee. And thou shalt not reveal it to the man who is not able to bear it, or to keep guard over this Book, but [only] to the wise who believe on Me, and who walk in My commandments. And whosoever hath gotten possession of this book, shall neither descend into the place of torment nor into Sî'ôl (Sheol). And moreover, whosoever shall carry it, and whosoever shall attach (or hang) it to his neck (or body) [Fol. 4*b*], his sins shall be remitted to him. And if he repeateth it with his voice at the time of the Offering (*i.e.* at the Eucharist), [his sins] shall be remitted to him, and he shall be cleansed from the pollution of sin. And if they (*i.e.* the priests) shall make at the bier (or tomb) the sign of the seal of SOLOMON thrice with this book, after he is buried, the angels shall conduct him in through the gates of life. And they shall make him to arrive before God, and shall introduce him into the kingdom of heaven."

O bring thou me, thy servant STEPHEN, into the light of life and into the salvation which is everlasting!

[THE FIRST SPELL]

And when JESUS had made an end of [these words], He told MARY His names which were convenient for [procuring] life and salvation (or health) [Fol. *5a*]. And again He said, "Let men cry out and say, 'I take refuge in Thy names,'"

Berhânâ'êl	'Afreyôn	'Afnâtâ
Laḥan	'Ûrâ'êl	'Afûr
Masdeyôs	Lâhî	'Afkîr
Yâw	Kêdâ	Khîṭâ
Mâryôn	'Afrâtâw	'A'ô
'Amânû'êl	'Adnâ'êl	'Aḳbadîr
Badmâhîl		

In these Thy Names I, Thy servant STEPHEN, have taken refuge, so that Thou mayest have mercy upon me, and mayest show compassion upon me.

Kîrôs	Baṭrôkôs	Ṣabîn
Tâtîn	Patîn	Derpîḳâwî'âl
Kamerleyôs	Tenberânem	Kerâdeyôn
'Awergâ'êl	'Akôṭeyâ	Kared'êl

[Fol. *5b*]

Yâkêr	'Afkâ'êl	Saḳelkelyânôs
Tarkîyôs	Kuebâ'êl	'Arnâ'êl
Debâ'êl	'Alyôs	'Îrôs
Ḥanô	'Alfâ	'Îyâ'êyâ
Hîdâ	Yûdâ	'Ûdâ
'Adâ	Dâldâ	Ḥarî
Dûni	Lawalâdî	Kôbâ
'Alfâ	Nîyôdîḥarî	Deldâ
'A'ûhadîdleyâdî	Nedlekîn	Hehedûdî
'Awyân	Terên	Ṭâtâs
'Akhâzyôs	'Atyôs	Mâsyâs
Bâ'êl	'Ahûhâ'êl	'Awlôdel
Dân	'Alnâtîn	

I have taken refuge in these Thy Names so that Thou mayest have mercy upon me, and show compassion unto Thy servant STEPHEN.

And JESUS CHRIST said unto MARY, " The mercy of My heavenly Father is complete and perfect. And if [men] believe [Fol. 6*a*] in this my Name, He will judge (*i.e.* assign to) them life and salvation (health)."

May it happen to me thus, thy servant WALDA MICHAEL, for ever and ever. Amen.

THE SECOND SECTION

IN THE NAME OF THE FATHER, AND THE SON, AND THE HOLY GHOST, ONE GOD.

And JESUS wrote with His holy hands [the following Names] :—

Sîrônô	Panâk	Wîpîrôs
Farases	Nôrôs	Mas'amar
Yâwsêf	Refseyôs	'Alhîyôs
Mag'eyôs	'Êlnôs	Fapalnâ
'Êflôn	Yar'ayôs	Dîdmôs
Rapyôn	Ḳuoḳuenafê	Yûsîf
Madfen	'Alfô	Maḳdeyôs
'Afrê	'Alîf	Bêt

[Fol. 6*b*]

Gâmêl	Dâlêṭ	Hê
Wâw	Zây	Ḥêt
Ṭêt	Yûd	Kâf
Lâmêd	Mîm	Nôn
Sâmkît	'Ê	Pê
Ṣadê	Ḳôf	Rês
Sân	Tâw	

May my petition draw nigh unto Thee, O Lord! By the might of these Thy Names, let not one make me to see the smoke of the place of torment, Thy servant STEPHEN.

O Great God, Who endurest for ever, what are the Names which our Lord told the divine PETER? Here are they :—

Fekîyer	Lâhû	Mesdeyâs
'Aṭen	'Aflâ	'Alên
'Aṭlâḳîn	Lâhlâhû	La'enâḫanaṭû(?)
Neḥlef	'Aryôs	Waryôs
'Aḳleyâ	Pelyâ	Tashîhâlô

[Fol. 7*a*]

Mîd	Ḥa'ê	'Ayô
Remâkermîr	Sûryâl	Sadâḳâ'êl
Salâtyâl	'Afkeyâl	'Anyâl
Mîlmâ'êl	'Aṭyôd-'ay-lesân	'Alfâwî
'A'a-dakhârâwî	Yâw	'Agyôs
Kâfû	'Armenyâl	Semyâl
'Afrû	'Arânât	'Afrâskares
'Aihi	'Êlôhê	'Afmîyâl
'Amânû'êl	'Abresteyâl	'Alyâl
'Êrnâ'êl	'Amâseryâl	'Afseryâl
Germelyôl	Dermelyûl	Ḳardalyûl
Germûlyûl	Der'aswîs	'Arkeyâl
Sarseyasel	'Amyôs	Ṭêbêryâ
Hêtyô	Ṭersedem	Maryâ

[Fol. 7*b*]

Mârmâ	'Ansôs	Dâkê
'Abyâtêr	Ḥarâṭôn	Pankatarsâṭer
'Îyâsyonrôdakh	Khêdrâ	'Û'usûsinôyâk'a'eyôwôs
Salâs'êl	Hêsêwôn	Denpas

In the might of these Thy Names I, Thy servant, WALDA MICHAEL take refuge.

SÂDÔR 'ALÂDÔR DÂNÂT 'ADÊRÂ RÔDÂS.

In the five nails of the Cross of our Lord JESUS CHRIST, I Thy servant STEPHEN take refuge. [Here follows a cross.]

The Third Section

In the Name of the Father, and the Son, and the Holy Ghost, One God.

[Fol. 8*a*.] A prayer (*i.e.* spell) concerning the carrying of the deceased:

Deḳâs Batrôn Kûgûyâ, Gânôn, Kâwes Ḳîrel.

Nothing shall attack the dead body for whom this writing (or book) shall be recited on the last day.

On the day of the judgment of Gog and Magog, those who have defiled the Law of God, and those who bring forward corrupt speech, shall say, " I am Christ, the Son of the living God," and all those who are sinners will believe him (or them) [Then] Christian folk shall say, " We believe in the Name of Jesus Christ, in the Son of God, in the Father, and in the Son, and in the Holy Ghost. And Elias shall preach unto all [Fol. 8*b*] Christian people, and they shall believe in Christ, the Son [of God].

And whosoever believeth in the son of Satan (Antichrist ?) shall be condemned to punishment in the place of torment. And whosoever believeth in Jesus Christ, the Son of God, shall never enter the place of torment; he shall be held worthy and shall walk in (or with) the Holy Ghost. God saith, " I am the God of the heavens and the carth.

The king Nâtnâ'êl (Nathaniel ?) himself shall go about; the Christian shall lament (?) the tunic, the fountain of glory and life. This is he who shall ride the horses of life on the day of rewards and judgment. And on that day the sun shall become black, and the moon shall become blood.

In that day show [Fol. 9*a*] mercy and have compassion upon me, thy servant Walda Michael.

Praise be to the Father, to GOD in the heavens, and peace on the earth! He who hath separated the light [from the darkness], our GOD and SAVIOUR, shall instruct us, we making mention of Thy Name, and supporting ourselves on Thy Cross. And we place our confidence in Thy hidden Name, I will give praise unto Thee among the young and the aged, so that Thou mayest show mercy, and mayest have compassion upon me, Thy servant STEPHEN.

And the Angel of GOD said unto Him, " What now is this noise of thunder which I hear? " And GOD said unto MICHAEL, " It is that which cometh from the place of torment, which is the habitation of sinners, and of those who have not performed the Will [Fol. 9*b*] of My Father; [it ariseth through] the destruction of the souls of those who have treated His word with contempt."

And He said unto our Fathers, " On the day of the [bestowal] of rewards, and [the assignment] of punishment, blessed shall be the man who hath had this book written for him, and blessed shall be the man who hath suspended it from his neck, and hath placed his confidence therein, for GAHÂNAM (GEHENNA) shall never seize him. And in that day there shall be a sun that shall never set, and a lamp that shall never be extinguished, and the sound (or voice) of their reward which shall never be silenced, and praise of His kingdom which shall never be rooted out, and His throne crowned with fire, four [fold ?], which shall never, never be overthrown, Amen."

And His angels said unto Him, " Recite to us Thy Name, so that we may praise Thee, and sing [Fol. 10*a*] hymns unto Thee. " And GOD said unto them :

" My first Name is 'ĪYÂWÂDÂ.
My second Name is KÎNYÂ (ARTIFICER).
My third Name is 'AMÂNÛ'ÊL (EMMANUEL).
My fourth Name is 'ĪYÂSÛS (JESUS).
My fifth Name is KERESTÔS (CHRIST).
My sixth Name is 'ĪYÂD.
My seventh Name is 'ĔGZÎ'ABEḤÊR (LANDLORD).

If there be a man who hath placed his confidence in these names, and who hath performed a ceremony of commemoration of me, I will show him mercy [and will save him] from this devouring fire, and the worm that never sleepeth, and the fire which is never extinguished, and the smoke which never dieth down."

And GOD said unto MICHAEL, " I have given unto thee the power of bringing offerings of praise unto me. If there be any man who hath performed a ceremony of commemoration of me, and who hath put his trust in me, and hath suspended this Book [from his neck] and [Fol. 10*b*] carried [*i.e.* worn] it, or laid it up in his house, and if he hath in his firm faith drunk the water of his prayer, the torment of hell shall not draw nigh unto him."

And straightway MICHAEL the Archangel bowed low and made obeisance to GOD. And he said unto them [*i.e.* the angels], " I give thanks unto the Lord my GOD, Who hath made me to see the marvellous thing which shall be performed at the last day."

And then all his [*i.e.* MICHAEL'S] angels gathered themselves together that they might have that Book read [to them] by CHRIST, the Son of GOD. Now that Book had been sealed with the Seal of the Father, and the Son, and the Holy Ghost, and no one had the power [or was authorized] to open that Book, except the Four-and-Twenty Priests [Fol. 11*a*] of heaven, and the Four Evangelists. And the Four Evangelists took that Book, and they opened the seal thereof, and they looked therein, and they read it out aloud so that [the angels] might hear. And straightway the angels took seven trumpets and blew blasts on them. And they took seven vessels and poured them out on the face of the earth so that the children of the good and righteous folk might be sanctified, and that they might be free of the heavens and the earth, and the Seven Gates, and the Seven Luminaries [or Regions of Light], and the Seven Bearers of the Throne of GOD. By this His Awful Name was made known to the Prophets and to the Apostles, in the places where they were, and in the Holy Mountain. On the sixteenth day of the month of Maskaram, at the sanctification of her body in purity through the honourable Cross of [Fol. 11*b*] CHRIST, and the tomb of our Lord JESUS CHRIST. He made mercy to appear on us, according to His holy word.[1] And He declared unto His saints, with honourable . . ., the Word of God.

'Agfôrâ	Zemrâ'êl	Gerkâ'êl
Demnâ'êl	Kîdû	'Adenâ'êl
Khîrût	Zebdeyôs	'Êmônyôs
Mîltârâ	Târbôtâ	Kamayâter
Nefyânôs	'Afôrâ	Nefyâd
Ḳatâwîr	Waryâ'êl	'Aldân, his name.
'Atawâs	Sasôrô	

Thus is the interpretation thereof in Gĕ'ĕz (Ethiopic).

[1] This passage is difficult, and some words seem to have been omitted by the scribe. On the sixteenth day of Maskaram the festival of the discovery of the Cross by Queen HELENA, and the consecration of the church and Temple and tomb of CHRIST was celebrated.

And because of the ascension of MARY into heaven, do Thou make to ascend also into heaven Thy servant WALDA MICHAEL.

[Here follow nearly three columns of names, each of which is prefixed by " one (numeral) His name "; here I give only the names.]

[Fol. 12*a*.]

Sâfyôs	Ḳôhôkî	Gabre'êl
Berhânâ'êl	Ṣerâ'êl	Zemrâdâ'êl
Dedyâ		

These names do not exist in the heart of mortal men; these are they (literally " this is that ") which came forth from the mouths and from the words of the lips of the Father, and the Son, and the Holy Ghost.

'Agyôs	'Arehnôn	Batrôn
'Asrârôn	Ṣenû'e	Mekyâr
Medyôs	'Agyôs	Maftelḥêm
'Ĕlmaken	[Fol. 12*b*]	'Ĕyâ
Mekyâr	Gânôn	Nadâdîhâ-lanafes
'Adâhêl	Gem'adyôs	'Agateyôr
Kedyôrôs		

These names are the keepers of the soul and the gates thereof, and he who beareth them in humility (or simplicity), and in silence, and in patience, and in humbleness of speech, and the fear of GOD, shall be saved [from the consuming fire].

The man who is willing to hear with his ear this word shall traffick in gold, and in silver, and in the apparel of honour; but he who shall fail to do so shall become a slave of DELESḲEYÂM (?). My Name here (or in this world?) is that with which Abbâ JOHN baptized me. On this day [Fol. 13*a*], and in this hour, it will open the gates of righteousness.

And [the man who knoweth it] shall not see the place of torment, and his work shall [not] be in the place of torment, and God shall show mercy upon him.

And because of the Ascension of MARY into heaven, even so do thou make me to ascend into heaven. In these Thy Names I have taken refuge, I Thy servant STEPHEN.

Yalô'êl	Sedeb'êl	'Îyô'êl
Fenô'êl	'Aḳna'êl	'Îyôbed
Ḳirôlôlâ'êl	'Îlîṣal	Salâtî'êl
'Êzrâ'êl	Ḳâlâtalâ'êl	'Azrâwî
'Êlâwî	'Êlâ'îrûbâlâ'êl	Sedrâ'êl
Sanbâ'êl		

Through these Thy names let neither death nor suffering come unto me.

Delâ'êl	Lek'êl	Felâ'êl

[Fol. 13*b*]

'Îk'êl	Dûlâfû'êl	'Îyâ'êl
Dereslâ'êl		

In all these Thy Names I, Thy servant STEPHEN, have taken refuge.

'Êlsâ'êlkôs	Pentâkôrôṭîs	'Agmîmûs
Ṭenten	'A'edân	'Akmâtûs
'Îyân'êl	'Azâ'êlḥagômâ	Marmôtônâgê
'Adêrâṣbeyôn		

By these Thy Names, and by the shedding of the blood of Thy servant GEORGE, remember me, O Lord, in Thy kingdom. Thy servant WALDA MICHAEL.

SÂDÔR 'ALÂDÔR DÂNÂT 'ADÎRÂ RÔDÔS

By the five nails of the Cross of our Lord JESUS CHRIST [remember me] for ever and ever. Amen.

[The Fourth Section]

[Fol. 14*a*.] In the Name of the Father, and the Son, and the Holy Ghost, One God.

Hearken, O our brethren, and we will speak unto you. Peradventure ye will believe the word of Lêfâfa Ṣedeḳ, which God gave to Mary as a covenant for the last day—now, she bore these names in her womb after the manner of a helmet (*i.e.* a protective covering)—whosoever beareth (*i.e.* carrieth) these names like Mary, or tieth this book [to his neck, or body], [Fol. 14*b*] shall never see the place of torment, but shall find life everlasting.

And our Lady Mary asked our Lord and said unto Him, " Tell me which is the greatest of all these Names of thine." And our Lord Jesus Christ answered and said unto Mary, " I will tell thee these my Names which, though difficult for the hearing, and are hidden from the sight, are beneficial to him that is able to bear them and to keep them safely."

And again our Lady Mary said unto Him, " I beseech Thee, O my Son, to tell me Thy hidden (or secret) Name."

And our Lord said unto her, " I will tell thee my Name correctly and thou shalt not hold these my Names lightly, for it is a difficult one for the man who is not a believer by nature. And as for the man [Fol. 15*a*] who is unable to bear this my word, it is not seemly to reveal my Names to him."

And again our Lady Mary asked Jesus, and said unto Him, " I will not declare them unto foolish men, or unto those who have no understanding in their hearts, or unto those who do not seek the habitation which is in the heavens, or those who have not rejected the honours which belong to this earth."

And our Lord answered and said unto MARY, "Separate (?) not thyself. [I would] that men should know my Names which I will tell thee."

And having made an end of speaking, JESUS stood on a pillar of cloud, and He appeared to MARY in a flame of fire until He had declared unto her all these Names. And He said unto her:

'Êlôhê	'Êlôhê	'Êlôhê
'Êrân	'Êrân	'Êrân
Râfôn [Fol. 15*b*]	Râfôn	Râfôn

And this is interpreted 'Akhâzî 'Âlam (Sustainer of the world), Kasâfî, which is Maḥarî (Merciful One), which is Maryôn, which is 'Îyetma'â'e (Cannot be provoked to wrath), which is Fôfôrân, which is Tashâhâlanî (Have compassion on me), which is Beyôn, which is Khêr, which is Baresbâhil—every one [of which] a man shall fear.

The Name of the Father is Mâryâl.
The Name of the Son is Menâtêr.
The Name of the Holy Ghost is 'Abyâtêr.

In these Thy Names I take refuge. I Thy servant STEPHEN.

[O] Mîkâ'êl (Michael)
and Gabre'êl (Gabriel)
and Sûrûfêl (Seraphim)
and Kîrûbêl (Cherubim)
and Suryâl (Suriel)
and Rûfâ'êl (Raphael)
and 'Îyâ'êl
and Sâḳû'el

ye Seven (*sic*) Archangels make supplication for us, and make intercession on our behalf.

[O] Sadûḳâ'êl
[O] Bernâ'êl

make ye supplication [Fol. 16*a*] on our behalf in your prayers so that we may be saved.

'Egra-mâtâ
Surteyôn
Marâmârâ
Malîṭôn
} The Four Beasts.

I take refuge in thy Names.

'Aldân Thy Throne, and
Lemḥesâ Thy City, and
The highest heights of Thy habitation, and
MARY, who gave Thee birth, and
The Four Evangelists, and
The Fifteen Prophets, and
The Twelve Apostles, and
The Twenty-four Priests of heaven, and
The Forty Soldiers of heaven, and
The Seventy-two Disciples, and
The Five Hundred Friends (?), and
The Three Hundred and Eighteen orthodox [Fathers], and
The Seven Archangels, and
The tens of thousands [of angels].

In the Names of these and in the Names of all the holy angels, I thy servant WALDA MICHAEL have taken refuge [Fol. 16*b*].

SÂDÔR 'ALÂDÔR DÂNÂT 'ADÊRÂ RÔDÔS.

In the five nails of the Cross of our Lord JESUS CHRIST, I Thy servant STEPHEN have taken refuge. [Here follows the drawing of a cross.]

[The Fifth Section]

In the Name of the Father, and the Son, and the Holy Ghost, One God.

The Prayer (*i.e.* Spell) of the journey to heaven through the . . . of the earth.

Protect Thou me, O Christ, so that the angels of darkness may not obstruct my soul. And let there be sent unto me the angels of light. Michael and Gabriel—those august angels—and the Paraclete, and the Spirit of [Fol. 17*a*] Righteousness, so that the angels of darkness may never obstruct my soul, and that the Lord may not make me to stand in the darkness [amid] the gnashing of teeth.

I, Thy servant Walda Michael take refuge in Thy Name "Genpâwê"; and in the Name of Mary, the Virgin, the God-bearer, Ṭebreyâdôs (*sic*); and in the divinity of the heavenly beings and the heaven of heavens; and in the Throne of the praise of Him Who hath builded His citadel. There is none in whom a man may believe except Christ, the Son [of God], the Merciful. Say thou unto me, "I have shown mercy unto thee, forgive Thou the sins of me, Thy servant, Stephen."

And in the world which is to come hereafter, and in this world also, let the Seven [Arch]angels, and the Seven Pavilions take and [Fol. 17*b*] carry up the prayer on behalf of men for mercy, O Thou Who art the Lamp of His angels.

Then one saith unto Him, "Lord, who is the man that hath not transgressed? Which wood (or tree) is it that will not give forth smoke? [Among] the sons of men, who is the man that hath not committed sin? There is none good except Thee."

And straightway God spake unto the Twelve Apostles, and to the Seventy-two Disciples, and commanded them to write copies of this Book. And He said unto His Apostles, " I give you permission and ye shall recite it to every one who believeth on me, [and] in the Name of Jesus Christ, the Son of God, Blessed is he who shall believe in me. Whosoever shall write down the word of this Book [Fol. 18*a*], and he who shall have a copy of it made, and he who shall hang it about his neck, having washed himself in the water of prayer, and he who shall lay the Book up in his house, shall never die the death. And they shall live at the last day, and on the day of judgment and punishment mercy shall be shown to them. And I will spare the fire of Gehenna, on the day when sinners and transgressors are separated [from the righteous punishment]. The man who carrieth this Book, wheresoever he may be, whether by day or by night, blessed shall he be."

By this Book, which driveth away devils, and beareth away (?) death from over the soul of Thy servant Walda Michael, the devils are also driven away from over the Throne of the praise of God [Fol. 18*b*] for ever and ever. Amen.

In Demâhîl, the Name of Thy might,
And in Tôbîl Thy name,
And in Leḳ'êl, Thy baptismal Name,
In Guôhûkâ'êl, whereby Thou didst burst open the mansions (or citadels) of Sheol,
In Ḳatanâwî, and
In Satanâwî, and
In Ḳarnalâwî, Thy Name.

I take refuge [in these] so that Thou mayest have

mercy upon me, and show compassion upon me, Thy servant, STEPHEN.

Thou Who wast crucified, the son of MARY, the Nazarene, the King of JUDAH, remember me, O Lord, in Thy Kingdom. Thy servant WALDA MICHAEL.

SÂDÔR 'ALÂDÔR DÂNÂT 'ADERÂ RÔDÂS.

By the five nails of the Cross of our Lord JESUS CHRIST, and in these Thy Names I have taken refuge, and I have made both my soul and my body to have refuge therein [Fol. 19*a*], I Thy servant STEPHEN, for ever and ever. Amen.

[THE SIXTH SECTION]

IN THE NAME OF THE FATHER, AND THE SON, AND THE HOLY GHOST, ONE GOD.

The BOOK OF THE DISCIPLES, who asked and entreated JESUS until He revealed unto them His hidden (or secret) Name.

And after this (*i.e.* their entreaty) JESUS spake to them, and said unto them, "Guard ye it (*i.e.* the Book), and make it to endure, and ye shall be saved from the fire. Whosoever shall take heed to know my names, and whosoever [Fol. 19*b*] shall make it to endure, and shall recite it, and whosoever shall cause it to be read, having washed himself [in the water of prayer, shall be saved from the multitude of his sins."

This is the [Book] which our GOD [spake] with His voice, and wrote with His holy hands, and gave to His disciples that they might read it, and in the reading thereof they found His Name, and they rejoiced and were glad. And they said, "Thanksgiving and praise be to Thy Name, O Thou Who hast shown us all this, and hast given unto us Thy holy Name." And they cried out (or proclaimed) His Name, saying,

	Râfôn	Râfôn	Râfôn
	Râkôn	Râkôn	Râkôn
	Pîs	Pîs	Pîs
	'Aflîs	'Aflîs	'Aflîs
	Melyôs	Melyôs	Melyôs
	Ḥanâ'êl	Ḥanâ'êl	Ḥanâ'êl
	Ṣerâ'êl	Ṣerâ'êl	Ṣerâ'êl
	Nârôs	Nârôs	Nârôs
[Fol. 20*a*]	Kîrôs	Kîrôs	Kîrôs
	Fêlôs	Fêlôs	Fêlôs
	Sîrôs	Sîrôs	Sîrôs
	Lîfernâs	Lîfernâs	Lîfernâs
	Nîrôn	Nîrôn	Nîrôn
	'Îrôn	'Îrôn	'Îrôn.

Of all my Names those which are the greatest are :—

Demâhîl	Demâhîl	Demâhîl
Beresbâhîl	Beresbâhîl	Beresbâhîl
'Aḳmâhîl	'Aḳmâhîl	'Aḳmâhîl.

There are none who know this my Name except the Four-and-twenty priests of heaven, and MARY, my mother.

And Jesus said unto them, " By this my Name ye shall be saved, and your sins shall be remitted unto you. And as with you even so shall it be with him that keepeth it, and doth believe. He shall be saved, and he shall not be put to shame before me, and he shall not see the smoke [Fol. 20*b*]. Of all the prayers (*i.e.* spells) which are written in this my Book, there is no formula greater than this. Whosoever believeth in this [prayer] I swear by my throne, and by my exalted head, and by the stool which is under my feet, and by MARY, my mother, that I will show mercy unto him. This I swear by my holy angels and I will neither do violence to my righteousness, nor will I make my word to be a lie. And I will not befoul my covenant."

As Thou didst save the saints Thy Apostles, even so save Thou me by the might of Thy holy name; wash Thou me and cleanse me from my sin, me, Thy servant WALDA MICHAEL.

And again JESUS said unto them [Fol. 21*a*]:

"Blessed is the man who hath read (or had read), this prayer.

Blessed is the man who hath washed himself in the water of prayer.

Blessed is the man who hath heard this prayer with his ear.

His strength shall be like the strength of the rock.

He shall hear the sound thereof as if it were the roaring of a lion.

And I myself will protect him with my own might and strength.

And I will love him as if he were my disciple.

Blessed is the man who shall bear (*i.e.* wear) this prayer.

No unclean spirits shall draw nigh unto him.

Nothing shall be able to disturb the body and the soul of the man who hath this prayer with him.

Neither pain, nor weariness, nor hunger shall enter his house.

And he shall be able to drive away even SATAN, who shall not be able to draw nigh to his habitation.

And the thief [Fol. 21*b*] shall not be able to steal from him, and his foe shall not be able to overpower him; and he shall be able to exhaust the strength of every enemy of his.

And his house and his children shall be blessed.

And the angels shall never be far away from him.

The blessing of the Prophets and the Apostles, and the Spirit of GOD shall rest upon him at all times.

And the Spirit of SATAN shall be remote from him."

[Address to the reader]

And as for thee, if thou believest in this prayer, the water of prayer shall not be poured out into the earth. For it is honourable and holy, and is like unto the Body and Blood of CHRIST. It is a cleanser of sin, and a medicament of salvation for the soul and the body.

And when thou hast recited this [prayer], having washed thyself [in the water of prayer], thou shalt vanquish and overcome thine enemy and thy foe [Fol. 22*a*]. And no one shall be able to stand before thy face; all created things shall tremble before thee, and as soon as they see thy face they shall take to flight. And thy speech shall be grateful unto every man.

O my Lord, when Thou comest unto Thy kingdom remember Thy servant STEPHEN.

[THE SEVENTH SECTION]

IN THE NAME OF THE FATHER, AND THE SON, AND THE HOLY GHOST, ONE GOD.

THESE ARE THE NAMES WHICH OUR LORD TOLD SAINT ANDREW THE APOSTLE.

And [JESUS] said unto him, " Go thou to the city the eater of men (*i.e.* the Cannibal City) wherein is thy brother MÂTYÂS, that thou mayest bring him out of the prison house. Rise up and depart with two of thy disciples." And ANDREW answered [and said], "How is it possible for me to come to that city? For it is a very long way off [Fol. 22*b*], a journey of two years. It is impossible for me to get there forthwith, for there is a great sea [between that city and this place]." And the Lord answered and said unto him, "Fear thou not, O ANDREW, my beloved. I will reveal unto thee a formula which is great, and I will tell thee therein [my] Names. When thou arrivest and art ready to march [into the city], O ANDREW, say thus:

'Aryâsyâsnôs	'Aryâsyâsnôs	'Aryâsyâsnôs
Kîyâyûdûyôs	Kîyâyûdûyôs	Kîyâyûdûyôs
'Akleyâdâ'êl	'Akleyâdâ'êl	'Akleyâdâ'êl
Sarnû'êl	Sarnû'êl	Sarnû'êl
Tâdâ'ôs	Tâdâ'ôs	Tâdâ'ôs
Redyâ'êl	Redyâ'êl	Redyâ'êl

These were the Names of my Father before [Fol. 23*a*] we created the heaven and the earth. I will tell thee my Names, but first of all I had to tell thee His Names. To my heart belongeth my Name[s which are]:—

Salgâwâtâ'êl	Salgâwâtâ'êl	Salgâwâtâ'êl
Ṣabartnâ'êl	Ṣabartnâ'êl	Ṣabartnâ'êl
Tâdâ'êl	Tâdâ'êl	Tâdâ'êl
'Agesyâyôs	'Agesyâyôs	'Agesyâyôs
Lemyôs	Lemyôs	Lemyôs
'Astâdâḳôs	'Astâdâḳôs	'Astâdâḳôs

(which is interpreted JESUS CHRIST.)

Dûdûmîl	Dûdûmîl	Dûdûmîl
'Asḥal	'Asḥal	'Asḥal

The Names of the Holy Ghost [are] :—

Parâḳlîṭôs

'Arâdyâl	'Arâdyâl	'Arâdyâl
Dâ'êl	Dâ'êl	Dâ'êl.
'Êlôhî [Fol. 23*b*]	'Êlôhî 'Êlôhî	Ṣabâ'ôt 'Adônây
Geyôs	Geyôs	Geyôs
'Agyôs	'Agyôs	'Agyôs

(which is interpreted " Holy, Holy, Holy, GOD of Hosts, the Perfect One, Filler of the heavens and the earth; holiness is Thy praise (or glory)."

'Alkenât

(which is interpreted, "Hallelujah to the Father.
,, ,, Hallelujah to the Son.
,, ,, Hallelujah to the Holy Ghost.
,, ,, Praise [be] to the Father.
,, ,, Praise [be] to the Son.
,, ,, Praise [be] to the Holy Ghost.

These are they Who are ONE, at all times together, now and for ever, world without end. Amen."

I have told to no one this word (*i.e.* formula) except MARY, my mother, and I have revealed it unto thee. Pray ye in [Fol. 24*a*] these my Names, and the gates shall be opened, and those who are bound prisoners shall be set free. If a man beareth (*i.e.* weareth) these names, and tie them to his person, his portion shall be with PETER, the chief of the Apostles. The EVIL EYE shall not look upon him, and the might of the Enemy shall not draw nigh unto him. Neither the might of evil devils shall assault him, nor the might of foul spirits, and the Power of Darkness shall not be able to overcome him. [Here follow the Names thus :—

Gêrâden
Mîlôs
Gâdên
Satanâwî
Ḳatanâwî
Tankaram
Ḳatâlî
Mâhyâwî

O CHRIST, the Son of GOD and the Son of our Lady MARY, Who didst fetter BERYÂL, even so do Thou fetter my foes and my enemies.

Remember me, O Lord, by the might of these Thy names [Fol. 24*b*] when Thou comest into Thy kingdom —Thy servant WALDA MICHAEL.

[THE EIGHTH SECTION]

IN THE NAME OF THE FATHER, and THE SON, AND THE HOLY GHOST, ONE GOD.

THE NAMES OF OUR LORD JESUS CHRIST, SIDRÂLÂWÎ—that death may not come unto me except at my appointed time.

'Awlâkît	Derdâs	Nârôs
'Elôn	Dalfôgîn	Gâdên
Yôṭâ	Bîbakuolâdîn	Sîdrâḳâ'êl
Kîraḳîṭîn	Dôlôtôlôn	Zarûbâ'êl
Sefûfâ'êl	Dôlôhôlôhîn	Tôlakîn
Kafâzîn	Gâzên	Fûlâka'êl
'Alfâ'êl	Dârâtân	Zerâ'êl
Galmâlâwî	Galawdeyân	'Îyâfên
Kalâdîn	'Abdâwî	Menâsîlâwî

[Fol. 25*a*]

Selnôdes	Delâwî	Gôldâfôn
Ḳalâ'êl	Dafû'êl	Sedrâḳâ'êl
Sîlî		

[In these names] I take refuge, I, Thy servant STEPHEN.

IN THE NAME OF THE FATHER, AND THE SON, AND THE HOLY GHOST, ONE GOD.

'Alfâ and 'Ō (Omega)
'Alfâ 'Alfâ 'Alfâ 'Alfâ 'Alfâ 'Alfâ 'Alfâ
'Īyâ'êl 'Īyâ'êl 'Īyâ'êl 'Īyâ'êl
Īyâ'êl 'Īyâ'êl 'Īyâ'êl
Hîdâ'êl Hîdâ'êl Hîdâ'êl Hîdâ'êl
Hîdâ'êl Hîdâ'êl Hîdâ'êl
Yôdnâ'êl Yôdnâ'êl Yôdnâ'êl Yôdnâ'êl
Yôdnâ'el Yôdnâ'el Yôdnâ'êl
'Ūrnâ'êl 'Ūrnâ'êl 'Ūrnâ'êl 'Ūrnâ'êl
'Ūrnâ'êl 'Ūrnâ'êl 'Ūrnâ'êl
Hîrnâ'êl Hîrnâ'êl Hîrnâ'êl Hîrnâ'êl.

[Fol. 25*b*] Hîrnâ'êl Hîrnâ'el Hîrnâ'êl
'Amîs 'Amîs 'Amîs 'Amîs
'Amîs 'Amîs 'Amîs

Dâhdâ	Negdekînî	Hehdâdî
Serâyâsyâl	Suryâl	Fârdyâl
'Arâdyâl	Sadrâl	Mûdûyâl
'Adônây	Mâsyâs	'Amânû'êl
'Akoâr	Marâdyâl	'Arâdyâl
Kaf'êl	'As'al	'Afteyâl
'Armâyâl	'Aḳte'al	'Ares'al
'Akyâl	Fânû'êl	Ḳatîtyâl
Retyâl	'Alyâl	Tîtâ'ôl
Yûlyâl	Kartîyâl	Sabteyâl
Mîtâ'ôl	Mîrâ'ôl	'Aksîfâ'ôl

'Awketyâl	Bîtyâl	Fêwâl
Sarwâl	'Anwâl	Fîlala'ôl
'Akrestîyâl	'Absî'ôl	'Awnewâl
'Arne'êl	Wâtîr	Nâ'ûs
Tîrân'arnâs	Zarik'abeg	Termen
Yâ'asîkô	Mîsônkes	Mâtîr

[Fol. 26*a*]

Nâsâkîb	'Aksenûnîyôs	'Ûnâr
Barâkîyâs	Rûstîwôn	Dâkîyâs
'Ōrneyâs	Ṭerâs	Kînâs
'Abṣâlôn	'Ansekô	Mûd
Meṭôs	Mût	Ketonâ
Lî'aṣâ(?)	Kînâ	'Araṣâ(?)
'Anyôs	Sârdî	Kalâsîn
'Ûsûrân	Mîrâ'ak	Wârôka
Wardî'aka	'Aṣamâ'ôl	Kônâ'al
Dôrân	'Arnî	Mârîk
Lasanek	'Amîyôs	Dawra
Bared	Meyâl	Mâsidenyâl
'Armeyâl	'Aryâmî	'Anâmyâl
'Aldyâl	'Awyâl	Yâ'áb
Fû'amâ	Fûyâmâ	Sardûr
Matawâdây	'Arâdyâl	Rawer
Fârûl	Ferteka	Sûhâl
Mîkâ'êl	Gabre'êl	Sûryâl
Sadâkyâl	Sarûtyâl	'Anânyâl
Rûfâ'êl	'Akhrâṭyâl	Khârmâsyâl
'Aḳmayyâl	'Afdâmyâl	'Arenyânyâl

[Fol. 26*b*]

'Asrâm	Zîdâ'ôl	Sûrûk
Mensûk	'Akhabreyânôs	Kîrûbêl
'Afnânyâl	'Atlewâ	Beresteyâl
'Abreyâl	'Abrâḳ	Rāg

Ferteyâl	Ferfâr	Fâmâwâwâl
Fânânyâl	Dîdyâl	Marâdkeyâl
'Afdekyâl		

O Holy Trinity, I, your servant WALDA MICHAEL, take refuge in each of your Names, and in the Names of your angels, and of your priests, so that the foul spirits and the hosts of Diabolos may not approach me on my right hand, or on my left hand, or before me, or behind me, wheresoever I may be.

'Îyâsyôn Rôdakhn Hedrâ
'Û'ûsûsînôyâkek 'Ayûwôs
Salâs'êl Hêsêwôn Dênpes

[I take refuge in these your Names, I your servant] STEPHEN.

APPENDIX

I.—THE VIRGIN MARY'S VISION OF HELL

[From Brit. Mus. MS. Orient. No. 605, Fol. 94*a*]

And JESUS answered and said unto me, " Come, let us go towards the west also, and I will show thee where the souls of sinners and the men of deceit live." And He took me up and carried me towards the west, and He brought me towards the boundaries of earth and of heaven. And I saw a large court wherein was no darkness, and it was filled with a river of fire. And I answered and said unto Him, " What is the explanation of this river, and who are they who dwell in this river ? " And He answered and said unto me, " These were not wholly cold, and they were not poor." And I looked there and I saw many people, men, women, young folks and children. And I looked again and I saw some men who were immersed [in fire] up to their breasts, and some who were immersed up to their lips, and some who were immersed up to their skulls.

And then I saw a great yawning abyss, and if a soul fell down to a depth of 50,000 cubits it would not reach the limit of that abyss. And I asked my Son, " For whom is this doom [decreed] ? " And He answered and said unto me, " For all those who have committed fornication, and those who have lain with men and beasts. . . . These are they who shall enter this abyss, and this shall be their punishment for ever."

And then I saw another punishment. I saw an old man, and fourteen cruel angels of darkness were carrying him along, and they brought him into the river of fire, and the river of fire engulfed him. And the angels set him upon a throne of fire, and the fire flared up about him as high as his breast and they hurled red-hot darts into his sides, and they poured fire over him from out of a vessel. And my Son said, " This shall be his punishment for ever."

And I looked again into that abyss and I saw a man whom the angels smote until he fell down on his face; and blood was pouring out of his mouth. And the angels cast him into the river of fire, which rose up about him to his breasts.

And further on in that river I saw a great and mighty man, and the angels of darkness seized him and plunged him into the fire. And they were pounding him with red-hot stones, and beating him with rods of lightning.

And I saw another man whom the angels were bringing along, and they were beating his face with rods of lightning and his face was blackened (or scorched) by the fire, and they plunged him into the river of fire up to his breast. And they were cutting out his tongue with a red-hot razor, and slicing off his nose with a red-hot sword.

And I also saw many other men who were scattered about like ashes in the fire, and they were suspended on pillars of fire. And I saw other men in the river of fire, and worms were gnawing them.

And then I looked and saw virgins who were arranged in darkness, and they had chains of fire tied about their necks, and the angels of darkness were dragging them into the river of fire.

And I saw people hanging from pillars of fire, and the angels of darkness were afflicting them, and panthers of fire were biting through their throats, and lions of fire were crushing their legs with their teeth.

And I saw others suspended on pillars of fire, and flames were driving them from one side to the other. In front of them were fruits of all kinds and fresh, sweet water, but the men could not reach either the fruits or the water. And I saw men blown about in the fire with their hands cut off, and in a pit filled with bitumen and sulphur I saw men with both their hands and their feet cut off.

And I saw a man whom four angels were holding suspended with ropes of fire; he was hanging head downwards with spears of fire penetrating his face. I also saw a pit of fire in which men and women were immersed up to their necks, and their breasts (or hearts) were filled with serpents and vipers, and four angels were driving red-hot spears into them.

And again I saw Gehenna, which was sealed with seven seals. And my Son cried out, saying, "Open ye the doors of Jahannam that MARY, my Mother, may see." And when those doors had been thrown open, and I had seen Jahannam, I was afraid and terrified, and I said unto Him, "What is this river?" And He said unto me, "Its name is Jahannam." And I answered and said unto Him, "Who are they who dwell therein?" He said, "These are they who said that the Son of GOD was not; this shall be their punishment for ever." And when all those who were under torture saw me, they cried out, saying, "Blessed art thou, O MARY and blessed is the Fruit of thy womb. Blessed are the eyes that see thee." And I answered and said unto my Son, "Have mercy upon them for my sake, O my Son; there is no man without sin." And my Son answered and said unto me, "From the Eve of the Sabbath until the dawn of the second day of the week, the sinners [in Jahannam] shall have respite from their torture. Be not sorrowful, my Mother. He who hath celebrated thy Commemoration, or hath called upon thy name, or who hath built a shrine to thee, or hath written a history of thy words, on him will I show mercy to the twelfth generation for thy sake, O thou who didst give me birth. This I swear unto thee by my Father, and by His Son, Myself, and by the flow of blood from my side, for the redemption of [the world], and by His Spirit, My Spirit." And when I heard His voice I gave thanks unto Him.

II.—MATTHIAS IN THE CITY OF THE CANNIBALS

[From the GADLA ḤAWÂRYÂT]

And it came to pass that when the Apostles were dividing the countries of the world [among them], and they were casting lots concerning them, the lot of MATTHIAS went forth that he should go into the City of the Cannibals (See LIPSIUS, *Apostelgeschichte*, Vol. I., p. 546 ff.). Now in that city they neither eat bread nor drink water, nor any other kind of food, but they feed upon the flesh and blood of men. And every traveller who cometh into that city the people seize and put out his eyes, and then they bind him in fetters until he hath lost his senses, and then they put him in a dark place, and feed him on herbs like an animal for forty days, and after that they bring him out and devour him. When MATTHIAS had come into the city, they seized him and blinded his eyes by means of a certain drug, with which they were acquainted, and they gave him grass to eat (now he would not eat thereof because the power of GOD was in him) and they cast him into prison.

[Having prayed for strength to submit to GOD's will in respect of him, GOD spake to him and told him that He would send ANDREW, the Apostle, to him, and that he would bring him out of prison.] And when MATTHIAS had been in the prison-house for seven and twenty days, our Lord appeared unto ANDREW when he was in the Country of the GREEKS, and said unto him, "Rise up and go unto MATTHIAS in the City of the Cannibals, that thou mayest bring him out from the prison-house, for the people of that city will in three days' time lead him out therefrom and eat him." And Andrew said unto Him, "If there be only three days left, I cannot come unto him; send Thine angel and let him bring MATTHIAS out quickly from the prison-house, for how can I get there in the next three days?" And our Lord answered and said unto ANDREW, "When the morrow hath come, do thou and thy two disciples rise up, and thou shalt find a ship ready to sail; embark therein, and it shall bring thee [to the City of the Cannibals]." And ANDREW rose up even as our Lord had commanded him. And came unto the sea-coast. Our Lord had made for him a beautiful ship, and He Himself was sitting there as the captain of it. And two angels were with Him in the forms of sailors. And ANDREW went to the ship and found our Lord sitting therein, and although he looked at Him he did not know Him to be our Lord. And he said unto Him, "Peace be unto Thee, O Captain of the ship"; and our Lord said unto him, "Peace, our Lord be with thee." [In answer to ANDREW's question our Lord told him that the ship was going to the City of the Cannibals, and when ANDREW said that he had no money to pay the fare, and no food, the Lord agreed to remit the fare and to feed him and his Companions. As ANDREW's companions were afraid of the sea, the Apostle sent them ashore, but at length they vanquished their fear and sailed with him. In due course ANDREW and his disciples composed themselves for sleep, and whilst they were slumbering, our Lord had them carried from the ship to the sea-shore].

Then ANDREW and his two disciples entered into the city, and there was none who saw them, and they came unto the gate of the prison-house wherein was MATTHIAS. And when they had laid hold upon the gate it opened unto them, and they went inside and found MATTHIAS sitting down and singing psalms, and they embraced him. And ANDREW said unto him, "Dost thou say, O MATTHIAS, that 'after the second day they will take me out, and slay me, and devour my flesh as if I were a beast.'" And MATTHIAS said, "I say that if it be the will of GOD that I come to an end in this city [it shall come to pass]." Then ANDREW looked at the men who were in the prison-house, and saw that they were bound like animals. And straightway he cursed SATAN and all his host, and he and MATTHIAS began to make supplication unto GOD, Who hearkened unto their petition. And they laid their hands upon the men who were in the prison-house, and their eyes were opened, and their minds returned unto them. And ANDREW commanded them to go out from the city, and he told them that they would find on their road a certain fig-tree, and that they were to sit down under it until the Apostles came to them. . . . Now the number of the men who went forth from the prison-house was one hundred and twenty and three. [When the men of the city went to the prison-house to bring out men to slay, they found the doors open, and the dead bodies of the seven keepers lying there, and they went back and reported the matter to the magistrates. The magistrates ordered all the old men of the city to be collected for slaughter, so that they might have food, and when the butchers were killing the victims ANDREW prayed to GOD, and the hands of the butchers withered. Then a severe persecution of MATTHIAS and ANDREW began, and the people seized them, and dragged them over the stones of the streets until their blood flowed, and then they cast them into prison and set a strong guard over them. The Apostles worked miracles and saved themselves from death, and they raised the dead who had been swallowed up in the flood which ANDREW brought upon the city. Then the people became converted, and ANDREW built a church for them, and administered to them the Holy Mysteries, and healed all the sick folk. The cannibals gave up their custom of eating human flesh, and ANDREW taught them to eat ordinary food. After this ANDREW remained in the city for seven days, and having seen them established in the fear of GOD, departed to the place whither the Holy Spirit led him.] A translation of the complete story of Andrew's visit to the City of Cannibals will be found in my *Contendings of the Apostles*, London, 1901, Vol. II., p. 267 ff.

III.—SAINT ANDREW AND THE DOG-FACE

[From the *Synaxarium*, Maskaram I.]

After this our Lord JESUS CHRIST commanded BARTHOLOMEW to go to the city of Barbar (*i.e.* the city of the Barbarians), and He sent to him ANDREW the Apostle with his helper to help him. And the people of that city were exceedingly wicked, and they would not receive the Apostles who were working before them signs and wonders. And GOD commanded one of the Dog-faced cannibals to submit to the Apostles, and not to resist them in anything which they ordered him to do, and they took him with them to that city. And the men of that city brought out wild beasts against the Apostles to eat them up, and straightway that Dog-face rose up against those wild beasts, and rent them asunder, and he also slew a great many of the men of that city. Because of this all the people were afraid; and they turned, and did homage at the feet of the Apostles, and they submitted to them, and they entered the faith of our Lord JESUS CHRIST—to Him be glory! And he appointed priests over them, and built churches for them, and the Apostles left them praising GOD.

IV.—THE PRAYER OF THE VIRGIN MARY ON BEHALF OF THE APOSTLE MATYÂS IN PARTHIA

[The first to describe this prayer was the great Ethiopic scholar LUDOLF (see his *Commentarius*, Frankfort, 1691, No. xxxv, p. 349 f.), who rendered the Ethiopic BÂRTÔS by BERYTUS, *i.e.*, BÊRÛT in Syria. This identification was accepted both by

DILLMANN and ZOTENBERG, but GUIDI has shown (see *Gli atti apocrifi degli apostoli*, Rome, 1889, p. 9) that BÂRTÔS is not Berytus, but PARTHOS, *i.e.*, PARTHIA, where, according to a tradition, one of the Twelve Apostles suffered martyrdom. And the apostle who was martyred in PARTHIA was not MATYÂS (MATHIAS) but St. MATTHEW. There is here a double confusion, one in the names of the apostles, and another in the names of the places. For the Ethiopic text of the Virgin's Prayer here translated see Brit. Mus. MS. Add. 16,245 (DILLMANN, *Catalogus, Codd. MSS. Orientalium*, Pars III, London, 1847, No. LXXVIII, p. 60), and Brit. Mus. MS. Oriental, No. 564 (see WRIGHT, *Catalogue of the Ethiopic MSS. in the British Museum*, London, 1877, No. CLXVIII, p. 112). A French translation was published by BASSET (*Les Apocryphes Ethiopiens*, Paris, 1895, No. V), who worked from MSS. 56 and 57 in the Bibliothèque Nationale in Paris. None of the texts are really good and the deficiencies in one have to be supplied from the others.]

IN THE NAME OF THE FATHER, AND THE SON, AND THE HOLY GHOST, ONE GOD.

THE PRAYER WHICH OUR LADY MARY PRAYED [ON BEHALF OF THE APOSTLE MATYÂS] IN THE COUNTRY OF PARTHIA. IT BROKE ALL [HIS] FETTERS AND DELIVERED THE DISCIPLE MATYÂS, [WHEREUPON] ALL THE PEOPLE OF THE CITY BECAME BELIEVERS. MAY HER BLESSING AND PRAYER BE WITH US! AMEN.

Our Lord and God and Redeemer, JESUS CHRIST—to Him be glory!—said unto His chaste Apostles and

His chosen and holy disciples concerning this prayer: "Among the angels of heaven there is none who knoweth it. The chiefs of the angels, the Cherubim and the Seraphim know it not, and there is among the celestial hosts none whatsoever who knoweth it, except the Father, and the Son, and the Holy Ghost, One God. My name is ''Alfâ' (Alpha), the first of letters; the name of my Father is 'Ala' (= Êl?), the counterpart thereof; and the name of the Holy Spirit is ''Arâdyâl.' Together we form One God, One Will, One Substance. I will teach thee, O MARY, my mother, to have the mastery over this great [knowledge] and to make thy petitions by means of this prayer."

When our Lady MARY heard these words of her beloved Son, she stood up on her feet, and she entreated our Redeemer JESUS CHRIST to rescue the Apostle Matyâs from [his] captivity, and to break forthwith all the fetters of iron wherewith he was bound in that city (*i.e.* BARTÔS). Then she turned towards the East, and lifting up her eyes towards the heavens above, as she stood by the side of her beloved Son, she began to make the following prayer: "Even thus do I make my petition unto Thee, my Lord and my God, my Son and my beloved One, my King, JESUS CHRIST. I am Thy mother MARY, I am MÂRIHÂM, I am the mother of the Life of the whole world, and I beseech and entreat Thee this day to hearken unto my prayer, and to send to me the angel hosts, the Seraphim and the Cherubim, and all the companies of angels, to carry out my plan, and fulfil my vow, and to do all the good things which I fain would do. For Thou art my hope and help, and in Thee do I place my confidence. My beloved Son, believing that this very day the stones (*i.e.* walls) shall be breached, that the chains of iron shall be melted, that the gates which are now shut fast shall be opened forthwith, and that the powers of darkness shall be made to flee away, and that all their strength shall disappear from every place wherein this [my] prayer shall be pronounced, I open my mouth and say :—

" Salutation to the Good Father Who sent greetings unto me by GABRIEL, the holy archangel.

Salutation to the throne of the Cherubim, whereon sitteth the ANCIENT OF DAYS (see Daniel vii. 11).

Salutation to the everlasting light which is upon Thy head.

Salutation to the Mighty Names, the Sixty Names of the Good Father, viz. :

'Alfâ 'Alfâ 'Alfâ 'Alfâ 'Alfâ 'Alfâ 'Alfâ
'Îyâ'êl 'Îyâ'êl 'Îyâ'êl 'Îyâ'êl 'Îyâ'êl 'Îyâ'êl 'Îyâ'êl
Hîdâ'êl Hîdâ'êl Hîdâ'êl Hîdâ'êl Hîdâ'êl Hîdâ'êl Hîdâ'êl
Yôdâ'êl Yôdâ'êl Yôdâ'êl Yôdâ'êl Yôdâ'êl Yôdâ'êl [Yôdâ'êl]
'Ûrnâ'êl 'Ûrnâ'êl 'Ûrnâ'êl 'Ûrnâ'êl 'Ûrnâ'êl 'Ûrnâ'êl 'Ûrnâ'êl
Hernâ'êl Hernâ'êl Hernâ'êl Hernâ'êl Hernâ'êl Hernâ'êl Hernâ'êl Hernâ'êl
'Ômîs 'Ômîs 'Ômîs 'Ômîs 'Ômîs 'Ômîs 'Ômîs 'Ômîs
Dehdî Neldikani Hehdûdî

[Some names have been omitted.]

Salutation to Thy holy resting-place (shrine ?).

Salutation to the veil of Thy shrine.

Salutation to the angels who sat with the Father when as yet He had not completed His work by means of the virgin to whom He sent GABRIEL to say, ' The Son of God shall come to thee.'

Salutation to thee, Mother of CHRIST, Who reigneth in peace.

Salutation to the Virginity which was not done away.

Salutation to the salutation which the Father spake to His Son.

Salutation to the glorious throne whereon He sitteth at the right hand of the Father.

Salutation to Him Who when He was on the wood of the Cross turned His head towards me, saying, 'My mother, go in peace.'

Salutation to the eyes which fixed themselves on JOHN, [saying] 'JOHN, take my mother into thy house.'

Salutation to Thy mouth which sucked milk from my breasts.

Salutation to the hands which formed ADAM.

Salutation to the feet which went into Paradise.

Salutation to the Word of the Father, JESUS CHRIST, Who is with peace.

Salutation to Him Who said unto me, 'Ask what thou wishest, O my mother, wish for what pleaseth thee by means of this prayer. Whosoever shall be healed by this prayer of severe sickness and illness shall have a firm faith therein; by means of it transgressors and sinners shall be directed and brought into the way of life; by means of it those who are fettered in the bonds of SATAN, and are in captivity to him, shall be set free; by means of it those who are afflicted shall be relieved, and all those who are suffering in the bonds of misfortune and oppression. When this prayer hath been recited over them they shall be relieved forthwith.' "

And when the Virgin, having said these words

rightly, turned to the right and to the left, she saw the angel GABRIEL standing there together with all the host of angels; and she was horribly afraid. And he said unto her, "Fear thou not, O MARY. I am the angel who bore to thee a message from the Father before thou didst bring forth thy beloved Son, Behold, I am come unto thee in order to fulfil thy request."

The Virgin answered and said, "My Lord, who is that holding a staff of gold in his hand?" And GABRIEL replied, "He is the archangel MICHAEL." Then she said unto him in a gentle voice, "I entreat thee, O MICHAEL, by my beloved Son Who hath committed unto thee power over all the angels; Who hath given unto Thee the rod of Command over the heavenly hosts which He hath taken from SATAN, the evil one; Who hath commanded thee to strip him of his glory and rank and power, and to hurl him and all his hosts into the depths of the abyss; Who hath handed over to thee the precious gifts of compassion and mercy so that thou mayest be able to intercede for every being; Who hath made thy name mighty and renowned. [I entreat thee] to fulfil all that I have spoken with my mouth.

And thou also, O GABRIEL, who didst bring me the announcement concerning the birth of my beloved Son, of whom when I saw thee, I was afraid, and said, 'How can this happen to me, for I know not man' (Luke i. 34). And when thou hadst heard my words, and didst know that I was terrified, thou didst answer and say, 'Fear not, O MARY; behold, thy cousin ELIZABETH hath conceived [a son] in her old age; and this is the sixth month with her, who was called barren. For with God there is nothing impossible' (Luke i. 36, 37).

Then I rose up and went into the hill country, and when I saw that Elizabeth had conceived I believed thy words, and I beseech and entreat thee this day to fulfil for me all that I wish.

I beseech Thee, O my beloved Son, by Thy marvellous birth, and by the words which Thou didst utter when I brought Thee forth in Bethlehem, saying, 'The name of the blessed Father is Fêlelemyô, the Name of the only-begotten Son is Tînô Tîḳânôs, and that of the Holy and Life-giving Spirit is Ḳuerḳuer-yânôs.' I demand this by the Five Nails which were driven into Thy body on the glorious Cross [their names] being

Sâdôr 'Alâdôr Dânât 'Adêrâ Rôdâs (see above, p. 37).

I beseech Thee by the Four Beasts (Rev. iv. 7, 8), who bear the throne of Thy majesty, and whose names are 'Alfâ (Bull), Lêwôn (Lion), Ḳuanâ and 'Ayâr, to send unto me the twelve armies of angels to remain with me until they have fulfilled every plan which is in my heart, and every word which is on my lips.

I beseech Thee, O Teryâl, by thy three secret names, that is to say, Danâs, Dîkî, and Marâfâ, and I will not let thee depart until thou hast fulfilled that which is in my heart.

I adjure you, O ye stars—thou Bêz, star of the morning (Lucifer), by thy great and secret name of Sûfâr, and by the power of the celestial beings who travel with you—whose names are 'Aksâr, Mardîâl, Madaryâl, 'Afeâl, 'Aseâl, and 'Aftîâl—I entreat and beseech you not to depart until ye have fulfilled that which is in my heart and mind.

I adjure thee likewise, O star of mystery of the evening, by thy great and mighty name of Sûrakîyâl, and by the powers who travel with you—whose names are 'Argâmyâl, 'Aḳtuâl, and 'Arsaâl—I entreat and beseech you not to forsake me until ye have fulfilled that which is in my heart, which I call upon you to perform.

I adjure thee, O Sun, by thy lasting name, by the power which is thine, and by all the power which God who created thee hath given unto thee, and by thy great light, and by thy chosen powers, whose names are Sûsaryâl, Fardîâl, 'Arayâl, Marâdyâl, Mardîâl, and I beseech and entreat thee not to forsake me until thou hast fulfilled the matter which I wish for.

I beseech thee, O Moon, who shinest in the night, and I beg and entreat thee by thy names and powers unto which thou hast been committed by God, and by the ordinances which thou obeyest, and by thy changes and revolutions, and I adjure thee by thy mighty name which hath been written down, not to leave me, and I entreat thee and thy powers not to depart before thou hast fulfilled all the wish which I have fashioned.

I command you, O Sun, and Moon, and all the powers that travel with you, that the Sun shall stand still at midday, and the Moon at midnight, until such time as my wish shall be fulfilled, and everything which at this moment is in my mind. May thy servant [So-and-so] live, being shielded from all suffering, and from every illness, whether external or internal, and that the favour of the Holy Spirit may rest upon him.

When ye shall go up to the Father, and when He shall question you, saying, 'Why have ye tarried this day, and why have ye not hastened to fulfil the duty which hath been assigned to you?' ye shall say unto Him, 'The Queen, the mother of the Lord GOD, the Creator, kept us back by adjuring us by Thy great and mighty and awful name, which none can resist uninjured; and she kept us back until we had fulfilled her vow and performed her wish.'

I adjure thee, O thou First Heaven, that wast created by the only-begotten Son in His wisdom, to unite thy endeavour to mine and to that of all the angels who are in thee, so that my wish may be fulfilled this day.

I beseech thee, O Second Heaven, by the wise 'ADÔNÂÎ, my beloved Son, who created thee by His word, to unite thy endeavour to mine for the fulfilment of my request.

I adjure thee, O Third Heaven, by the Truth Who fashioned thee, and Who hath established in thee the throne of His glory on which it resteth, I adjure thee by His awful Name, and by the throne of His glory, to fulfil my wish."

Having thus spoken, the Virgin lifted up her eyes and saw the heavens open, and she beheld her beloved Son seated on the right hand of the Father in the highest height of heaven. Then she turned round and saw the stones rend themselves asunder, and all the hosts of heaven appear, one after the other, beneath the throne of her beloved Son; when she saw this sight she made her prayer to her only-begotten Son. And at that same moment the iron [bolts] melted and flowed down like water, doors that were shut opened themselves, as did also the sepulchres and the tombs, the dead came forth, the devils were smitten with terror and fled, the earth quaked thrice on the right hand and on the left, and the Twelve Ranks of Angels came down from heaven, and followed after their captains, Then the Virgin said, " 'Adônâî, 'Adônâî, 'Adônâî, 'Amânûêl " (Emmanuel), CHRIST, my God, come quickly to me to fulfil that which I have in my mind."

Whosoever shall carry (*i.e.* wear) this prayer shall be delivered from sickness, and pain, and fever, and the calamities of a war with an enemy who is a foe to His servant. . . . Restore his body and his soul, and forgive him his sins, and let him be as [whole as he was] on the day of his birth. Let all evil spirits be remote from him, and let them all go back to their own place by the might of this prayer (*i.e.* spell) " Holy, holy, holy, is the Lord of hosts, Who filleth full the heavens and the earth " (Isaiah vi. 3; Rev. iv. 8) His glory is holy. MICHAEL is on His right hand, GABRIEL is on His left hand, RAPHAEL is before Him, SÛRYÂL is behind Him, SADÂKÎ'ÊL holdeth the crown over Him, SARÂTI'ÊL offereth to GOD praise and homage, and with him is associated 'ANÂNYÂL. I, MARY, entreat thee, and the hosts of light and ṢADAḲÎÂL, the angel of pity, to unite your endeavour to mine, so that Thy servant [So-and-so] may be healed of every sickness, both external and internal, and that his strength may be restored, and his sins, which Thou, O Lord, knowest, may be forgiven him. If it be Thy will, that this sickness shall persist in him for his punishment and benefit, do Thou send from heaven angels to help him, and to carry him to Thee without suffering, and may he experience Thy heavenly mercy both in this world and in that which is to come, for Thine is the power and the glory, and Thou art worshipped for ever and ever. Amen.

O ye Twenty-four sages of heaven, I adjure you by your names

'Akîâl	Fanû'êl	Ḳartîâl	Dartîâl
'Ilyâl	Zartîâl	Tîtâal	Yûlial
Kartîâl	Lebtîâl	Mîtâal	Mîrâal
'Aksitâal	'Awktîâl	Bîtâal	Râûâl
Sarûâl	Sakarûâl	'Anîûâl	Filâlêal
'Akerstîâl	'Aksîfâal	'Aûnûaâl	. . . .

I adjure you by the twenty-four crowns which are on your heads (Rev. iv. 4), and I beseech you to come and to make with your hands of light the sign of the Cross over this water and this oil, and not to depart until that which I have asked from the Lord and from you hath been performed. And moreover, I entreat you to fulfil this good work also by the name of my beloved Son, the Master of peace, and I adjure you by the Seven Veils which hide the Father, who is invisible. O ye Seven Angels who dwell by the Seven Veils, and whose names are :—

Bardâmîyâl	Wasîdênyâl	'Aremyâl	'Aryâmî
'Arnâmyâl	'Aldîâl	'Awyâl	

Ye shall not depart before ye have performed everything which is in my heart and mind, and for which I have asked you.

I adjure [each of] you, O ye Four Beasts who bear the glorious throne of the Lord, whose six wings are filled with eyes, two of whose wings cover his face, and two his feet, and with two he flieth. And ye proclaim the glory of God by day and by night ceaselessly (Isaiah vi. 2; Rev. iv. 8).

I adjure the three angels who sheltered me under their shadows when my Son was in my womb, and whose names are Yâab, Fâamâ and Fâyâm.

I beseech the three angels who protected my Son when He was in the stable wherein I placed Him in BETHLEHEM, and whose names are Sardûr, Matûadâî and 'Arâdyâl.

I beseech the three angels who protected the body of my Son, the only-begotten of the Father, when He was lying in the tomb, and whose names are Râûl, Fârûl and Fârtêkâ. I beseech and entreat you by the unrivalled majesty of the Father, and by the unimaginable glory of the Son, and by the grace of the Holy Spirit, Who proceedeth from the Father and emanateth from the Son, that ye may not be allowed to abide where ye are, until ye have all been to me and fulfilled all the good things for which I have wished.

I entreat thee, O Kâryân, thou star that didst shine when I gave birth to my beloved Son, and I command thee to shine in the face of him that shall carry this prayer. And if a man reciteth it in any place whatsoever, or over any person whatsoever, as soon as ever the evil spirits who are there shall see thy great light they shall flee. Protect thou the path of him that shall anoint himself with this oil, or shall drink of this water and this oil, or shall wash himself therewith, and disperse the darkness that would envelop him."

When MARY, the pure Virgin, had thus spoken, the earth quaked thrice, and she was afraid, and the angels came to her and said, "Amen, Amen, Amen." And our Lord sent MICHAEL the archangel to fulfil her wish, and He Himself spake unto her from the heights of heavens, with a sweet voice, saying, "O my mother, it is enough, for the earth quaketh. Thy prayer hath come unto me, even to the throne of my Father, the Creator of the universe; He, for my sake, will gladly fulfil all that thou hast asked for." And when MARY had heard these words, she ceased to speak, for she was astonished, and held her peace and said never another word. When she returned to her senses, she cried out with a loud voice, saying, " O Sûhâl, Sûhâl, shine thou upon me this day and until my wish shall be fulfilled."

And at that moment all the powers of heaven came to her, that is to say, MICHAEL, GABRIEL, RAPHAEL, SÛRYÂL (Suriel), SADÂḲYÂL (Zadkiel), SARÂTYÂL and 'ANÂNYÂL, the seven archangels. And they said unto her, " O Queen of all the women who are in the whole world, every thing which thou hast asked we will fulfil for thee, for thy prayer is mighty, gracious and efficient. "

And MARY answered and said unto them, " I would that ye would tarry here with me until I have recited this prayer :—

" Salutation to thee, GABRIEL, envoy of the King of the world.
Salutation to thee, MICHAEL, thou angel of salvation and mercy.
Salutation to thee, RAPHAEL, gracious and good, thou rejoicer of hearts.
Salutation to thee, SÛRYÂL, captain of the great host, friend of angels.
Salutation to thee, SADÂḲYÂL, comforter of the afflicted.
Salutation to thee, 'ANÂNYÂL, who together with the Twenty-four priests of the Spirit, dost present the prayers of the saints before the Lord.
Salutation to thee, SARÂTYÂL, who dost protect the souls of the saints and the righteous from the temptations of the devils who terrify the soul.

Salutation to thee, 'AHÊRÂTYÂL.
Salutation to thee, HERMÂSYÂL.
Salutation to thee, 'AFDÂMYÂL. [var. 'ARMYÂL].
Salutation to thee, 'ADSEMYÂL. [var. 'AḴMÂIYÂL].
Salutation to thee, 'ASRÂM.
Salutation to thee, ZIDÂÂL.
Salutation to thee, SÛRÛK.
Salutation to thee, MANSÛK.
Salutation to thee, ḤEBREYÂNÔS, the mighty Cherubim.
Salutation to thee, 'AFNÂNYÂL.
Salutation to thee, TÛÛÂL.
Salutation to thee, BARSTAL.
Salutation to thee, O my beloved Son.
Salutation to thee, O my King and my God, O CHRIST, who didst dwell for nine months and five days in my womb, and didst suck milk from my breasts, for Thou wast indeed man.
Salutation to all the angels who have come to me gladly."

Then all the hosts of heaven, and all the angels of the Spirit, and all those who stood round about the throne came to her, and they tarried with her until her prayer was ended and all that she wished was fulfilled. And the Virgin said unto them, "By the might of my beloved Son, go ye into every place wherein any man reciteth this prayer, and make ye the sign of the Cross over the water and the oil over which this prayer shall be recited in every country in the world."

And the hosts of heaven made answer, "Even so shall it be, O thou glory of all the women who are in the [whole] world, through the might of Thy beloved Son JESUS CHRIST, our Lord and our God." And the

Virgin said unto them, " By the might of my Son in this hour, and on this day, and world without end, whensoever a man shall anoint the body of thy servant [So-and-so] with this unguent, draw ye your swords of fire and drive away the evil spirits, and the malignant sicknesses which are in the body and limbs of the man who shall wash himself therewith, wheresoever and whensoever, and scatter them like the dust before the wind by the might of 'Abyâr, and 'Abrâḳ and Râḳ and Râdâ. I beseech you this day, O ye whose names are hidden, and who dwell nigh unto the Veil of the Father, to come unto me wheresoever I am, and fulfil the good wish which I have, for ye have not the power to disregard my wish. Satisfy ye every one who shall recite this prayer, whether he be in the East or in the West, in every district and in every country. O ye angels who dwell in the empyrean, Come ye to me this day, and heal him that beareth this prayer, and every one who shall recite it, and everyone who shall anoint himself in sincere faith with this oil and water. I adjure thee, O Heaven, to let the angels descend from the heights of heaven, and to come through this prayer, and to perform everything which I ask. Come to me, O ye Cherubim, who dwell in the heights of heaven, and heal every man who washeth himself with this water, and clotheth himself with this oil, Come O ye Four Angels, who stand on the four corners of the earth (Rev. vii. 1), and whose names are Fertiyâl, Ferfâi, Fâmûâl and Fanânyâl, I adjure you to come and fulfil my request. I adjure you, O ye Four Angels who guard the treasure-houses of the winds, and whose names are Didyâl, 'Afdâyâl, Dânâdyâl and Marâdekiyâl, dwell ye with me until ye have wholly fulfilled my request that Yâ'ḳôb may be healed of all

[his] sicknesses, both external and internal, I adjure thee, O Cherubim, who guardest the spring of the water of life in Paradise that none may drink therefrom, Come to me through this prayer, and heal thy servant [So-and-so]. Come thou to him, help him to do that which is good, renew both his soul and his body, pardon him his offences and sins, and make him to become even as he was when he entered the world."

When the pure mouth of the Virgin had uttered these words, our Lord JESUS CHRIST came to her in ineffable glory together with tens of thousands of tens of thousands of angels, and when she saw Him she said, " Blessed is he who cometh in Thy name! Thy coming this day is good, O my King, my God, my Son, Who didst tarry nine months in my womb." And the Lord said unto her, " Salutation to thee, O MARY, my mother. Verily I say unto thee that everything which thou askest on this earth I will fulfil. Whosoever maketh a petition in my name and in thine I will grant it to him, and everything which he may desire on this earth shall be fulfilled for him in the heavens. The wickedness of the devils who dwell under the earth shall be brought to nought, and when they hear the words of this prayer they shall fly from him, and they shall be dispersed like the smoke before the wind. And as wax melteth at the fire, even so shall the powers of darkness melt away wheresoever a man reciteth this prayer. Wheresoever are the names of my Father, and of thee, and of me, I will dwell every week, and my angels shall come every day, and shall bow down before the Father, and before my image and thy image. I am hearkening unto thee, O MARY, My mother, and I will do whatsoever pleaseth thee. Even as I chose thee, and held thee to be worthy for Me to dwell in so will I fulfil all thy petitions. Wheresoever shall be this prayer, there also shall be my blessing, and my grace, and my peace, and my love, and also shall be forever fertility, and abundance, and gladness, wheresoever it may be my angels shall protect the man [who reciteth it].

When a man shall recite this prayer in any place whatsoever, the spirit of evil shall not have power to draw nigh unto him, and it shall not be able to oppose him, for My name, and the name of the Father, and the name of the Holy Spirit and thy name, shall be found there, and shall remain there always. And they shall dwell upon him that writeth this prayer, and upon him that hath it, and upon him that receiveth it in faith and good will. And since thy mouth, O MARY, My mother, hath uttered this prayer, I will send MICHAEL and GABRIEL to every place where that prayer is to be found. And I have made it strong by My powerful hand, and by My life-giving Cross, and My mighty arm. I hearken unto thee, O MARY, My mother, and whosoever believeth in this prayer, and crieth unto Me in sincere faith, I will hear him and grant his request. Whosoever being ill, if he maketh entreaty unto Me in this prayer, I will heal him. I will go with those who are on a journey, and I will bring them safely to their homes; and if any man shall recite this prayer on behalf of those who are bound in prison, I will set them free. And if any man shall recite this prayer over the water and oil which are sprinkled over those who are possessed of devils, both the children of ADAM, and the offspring of animals shall be delivered. The cunning and wickedness of devils shall be powerless to harm him that carrieth this prayer, and the evil eye shall be unable to injure him. If this prayer be recited over a sick person, if he is to live I will heal him quickly, and if he is to die I, by means of the angels of light, will cause him to be carried to the place of light. I will bless the houses, and fields, and harvests of those who shall recite this prayer in sincerity and with a humble mind, and I will increase their crops, for there is no one who is like unto thee, O MARY, neither in the heavens nor on the earth. I assure thee that I will heal him that is sick, and if the spirit of evil lay snares for him I will deliver him by the might of this prayer, and I will restore him to the condition in which he was on the day of his birth. I will hear all those who beseech and entreat me in the words of thy prayer and request."

Even thus spoke our Lord and Saviour JESUS CHRIST, praise be unto Him and His mother the Virgin! Then He saluted MARY and went up into heaven in great glory.

V.—THE PRAYER WHICH THE VIRGIN MADE ON THE MOUNTAIN OF GOLGOTHA, WHICH IS THE TOMB OF OUR LORD ON THE 21ST DAY OF THE MONTH SANÊ (JUNE 26TH)

[For the Ethiopic text see Brit. Mus. MS. Harl. 5471 and Fol. 39 f. and Add. No. 16,233 (DILLMANN, *Catalogus*, Nos. LIII and LIX) and Brit. Mus. MS. Orient. No. 639 (WRIGHT, *Catalogue*, No. LXXXV, p. 52). For a French translation, see Basset, *op. cit.*, p. 11 f.]

"My Lord and my God, my Son and my King, JESUS CHRIST, Who of Thine own free will wast born of me, Who didst suck milk from my breasts, Whom the heavens cannot contain, Whom the bounds of the world cannot confine, Whom the earth cannot carry, Whose hand the space of the abyss, and the depths of the sea, and the rain-floods cannot fill, Whom the angels and the powers of heaven cannot draw nigh, my Son and my King, I, MARY Thy mother, Thy servant, beseech and make supplication to Thee. I carried Thee in my womb for nine months and five days. Thou hast dwelt in my body, Thou hast sucked milk from my breasts and hath lived upon my milk for three years, and I carried Thee on my back for five years. Remember, O Lord, that I have gone about with Thee for thirty years, and that I fled with Thee from one country to another when HEROD wanted to slay Thee. Hear my prayer, O my Lord, and my petition, O my Lord, my God. Remember, Lord, that I carried Thee in my womb for nine months and five days, and remember that Thou hast sojourned in my body. Remember, Lord, that I gave birth to Thee in Bethlehem during the season of ice and snow. Remember, Lord, that I left my country and went about with Thee from one country to another. Remember, Lord, my exile in a foreign land, and how I suffered hunger, and thirst, and wretchedness. Is there not reason for entreating Thee on behalf of the sinners as well as for the righteous who have celebrated my commemoration? My Lord, hearken to the prayer and petition which I set before me that Thou wouldst hear my words of entreaty and fulfil all that is in my heart this day [I beseech Thee] to send unto me forthwith twelve angels of mercy who shall tarry with me, and fulfil all that is in my heart, and the petitions for acts of grace which my lips make to Thee.

I beesech Thee, O my Son and beloved One by GOD, Thy Father, Who was with Thee before the creation of the world.

I beseech Thee, by CHRIST, Thy name which was with Thee before the creation of the heavens and the earth and of the angels and men, and of the sun, moon and stars, and before the night was separated from the day.

I beseech Thee by the Paraclete, the Holy Ghost, Who hath come forth from the Father, and Who proceedeth from Thee, and Who wast with the Father and the Son before the star of the evening, and the star of the morning made their appearance.

I beseech Thee by [my] womb wherein I carried Thee for nine months and five days.

I beseech Thee by [my] bosom, O my beloved Son, whereon Thou didst lie.

I beseech Thee, O my beloved Son, by [my] breasts which Thou didst suck for three years.

I beseech Thee by [my] back which hath carried Thee for five years.

I beseech Thee, O my beloved Son, by the hunger and thirst which I suffered for Thy sake when we fled from HEROD and went into the land of Egypt.

I beseech Thee by the tears which gushed from my eyes and fell on Thy glorious Flesh.

I beseech Thee by the mouth which kissed Thee.

I beseech Thee by the tongue which spake with Thee.

I beseech Thee by my ears which heard Thy gracious words.

I beseech Thee by my feet which walked with Thee for four and twenty years.

I beseech Thee by the bed on which Thou didst sleep.

I beseech Thee by the clothes wherein Thou was wrapped, O fire of the Deity.

I beseech Thee by MICHAEL, the angel of Thy wisdom.

I beseech Thee by GABRIEL, the envoy of Thy birth, who announced to me the glad tidings that I was to bear Thee.

I beseech Thee by RAPHAEL, the angel of mercy.

I beseech Thee by URIEL, the angel of protection and salvation.

I beseech Thee by ṢADÂKÎYÂL, the comforter of the sorrowful.

I beseech Thee by SALÂTYÂL, the righteous and just.

I beseech Thee by 'ANÂNYÂL, the angel of mercy.

I beseech Thee by the Four Beasts, each having six wings and many eyes, who bear Thy throne.

I beseech Thee by the Four-and-twenty sages of heaven who glorify Thee and burn incense before Thy throne.

I beseech Thee by the Ninety-nine Orders of angels who serve Thee.

I beseech Thee by the ten thousand angels on Thy right hand.

I beseech Thee by the ten thousand angels on Thy left hand.

I beseech Thee by the ten thousand angels who stand before Thee.

I beseech Thee by the ten thousand angels who stand behind Thee.

I beseech Thee by the tens of thousands of tens of thousands of angels who surround Thee.

I beseech Thee by the vast spaciousness of the heavens.

I beseech Thee by the great extent of the earth.

I beseech Thee by the angels of the clouds.

I beseech Thee by the angels of the sun and moon.

I beseech Thee by the angels of the hills and mountains, who were in being before their abodes were created.

I beseech Thee by the angels of fire.

I beseech Thee by the heavens, which are Thy throne.

I beseech Thee by the earth, Thy footstool.

I beseech Thee by Jerusalem, Thy city.

I beseech Thee by Mount Tabor, whereon were transfigured Thy Form and Similitude.

I beseech Thee by Mount Zion.

I beseech Thee by the Mount of Olives, the door of Thy kingdom.

I beseech Thee by JOHN, who baptized Thee.

I beseech Thee by Thy Holy Spirit.

I beseech Thee by Thy Holy Cross.

I beseech Thee by the nails [driven through] Thy hands and feet.

I beseech Thee by Thy holy Body and glorious Blood.

I beseech Thee by Thy Passion and Death.

I beseech Thee by Thy dwelling in the bowels of the earth for three days and three nights.

I beseech Thee by Thy entrance among the dead.

I beseech Thee by Thy descent into Sheol.

I beseech Thee by Thy Resurrection from the dead on the third day.

I beseech Thee by Thy Ascension into heaven with great glory.

I beseech Thee by Thy Second Coming.

I beseech Thee by the flame of Thy throne.

I beseech Thee by the exaltedness of Thy abode.

I beseech Thee by Thy years which never end.

I beseech Thee by Îyû'êl, Thy name which overcame the Enemy.

I beseech Thee by Thy name Tâdâ'êl, which the Enemy could not overcome.

I beseech Thee by Thy name Sêḳâ.

I beseech Thee by 'Ĕgzî'abeḥêr, Thy name before the creation of the world.

I beseech Thee by Thy hidden name, which cannot be uttered.

I beseech Thee by Thy revealed name, which is unknown (?).

I beseech Thee by SÂDÔR.

I beseech Thee by 'ALÂDÔR.

I beseech Thee by 'ADÊRÂ.

I beseech Thee by DÂNÂT.

I beseech Thee by RÔDÂS.

I beseech Thee by SIDÂ'ÊL.

I beseech Thee, O my beloved Son, to dwell with me, that the gates of the prisons may open of themselves, that the power of the devils may be removed from every place wherein they are, that the powers of darkness may be expelled, that the abodes of idols may become like water [courses], that all the temples of false gods may be laid waste, that their images may be broken in pieces, that all idols may be smashed, and that all the power of darkness may be destroyed. And I would that all the bonds of sin may be undone. And let all those who have had faith in this prayer be delivered from sin and set free by the voice of Thy heavenly Father, and by Thy saving voice, and by the voice of the Paraclete, the Holy Spirit, whose mouth (*sic*) is sharper than the razor (knife?), which separateth one root from another, and the soul from the body. O my beloved Son, I beseech and entreat Thee to hearken unto the words of my prayer, and to come with me, and fulfil everything which is in my heart."

When our Lady, the Virgin MARY, had thus spoken, the earth quaked, the rocks split asunder, the tombs revealed themselves, and the doors that were shut opened of themselves; and the Twelve Ranks of angels, following their captains, came down from heaven. And with them there came our Lord and Saviour JESUS CHRIST, Who had with Him ten thousand times ten thousand angels, ten thousand on His right hand, ten thousand on His left hand, ten thousand before Him, and ten thousand behind Him. There were Seven Lights before Him, and Seven Lights behind Him, and Fourteen Lights, which were brighter than ten thousand suns and moons, before His face.

At the sight of these our Lady MARY was seized with great fear, and she fell down upon the ground as one dead. Then our Lord and Saviour JESUS CHRIST stretched out His hand, and raising her up made her to stand before Him. And He said, "My mother, what hath happened? Why weepest thou? —thou who didst carry me in thy womb and on thy back. What hath frightened thee and terrified thee so greatly that thou hast fallen to the ground?" The blessed Virgin answered and said unto her beloved Son, "I have never before seen Thee thus. I who have carried Thee in a mortal body now see Thee [enveloped] in a mighty power of fire. Formerly when I saw Thee Thou hadst the form of a man, but now I see Thee having a terrifying and mighty appearance."

Our Lord answered and said unto her, "O my

mother, who didst carry me in thy womb for nine months and five days, who didst carry me on thy back and didst feed me with the milk of thy breasts, sweeter than honey and sugar, whiter than the milk [of other women], flowing more freely than the water of the Garden of Eden, what can I do for thee? For what work hast thou called me, O MARY, my mother? What petition can I grant? What can I do for thee?"

And the blessed Virgin said unto her beloved Son, "My beloved Son, my Lord JESUS CHRIST, my God, my Saviour, and my King! Thou art my hope, my asylum, my strength; in Thee do I put my trust. I was strengthened by Thee when I was in the womb of my mother, and Thou didst protect me therein, and of Thee will I make mention at all times and for ever. And Thou wast born of me by Thine own free-will, and with the permission of Thy Father and the Holy Spirit. Now, O Lord, hearken Thou unto my prayer and petition, and incline Thine ears to the words which my mouth shall utter. I am Thy mother and Thy servant, I beseech Thee to build indestructible habitations of light for those who shall celebrate commemorations of me, and shall build churches in my name. Do Thou array in the apparel of the heavenly marriage feast, and dress in the panoply of justice, which shall not wear out, which is fair to look upon and hath not been made by human hands, the man who shall clothe a naked man in my name. Visit with Thy mercy and compassion the man who shall visit the poor in my name. Set Thou at Thy heavenly table, O Lord, the man who shall feed him that is hungry and give drink to the thirsty in my name. Make Thou to drink of the river of the water of life which floweth in the Garden of Eden the man who shall nourish him that is famished in my name. Comfort Thou him that comforteth the suffering one in my name, and comfort him when his soul shall depart from his body. Lord, make to rejoice the man who cheereth him that is sad, and set him among all the saints who please Thee and fulfil Thy will. Write Thou in the Book of Life, with a pen of gold, the name of him that writeth this book, or who hath a copy thereof made. Bestow Thou, O Lord, upon the man who suspendeth this prayer from his neck a reward, the like of which the eye of man hath never seen, nor the ear of man hath never heard of, nor the heart of man hath ever imagined.

I beseech and entreat Thee, Lord, to deliver from Hell every one who believeth on me. Make him that shall sing my praises on the day of my festival, to hear the songs of the celestial choirs of angels."

And the Lord said unto her, "It shall be even as thou sayest. I will build habitations of light, and give a glorious seat in the kingdom of the heavens, and obtain the grace of my Father and the Holy Spirit for him that shall build a church dedicated to thee.

The man who shall visit the sick in thy name I will visit when he is sick and prostrate on his bed. When he departeth from this fleeting world I will not make him to drink of the bitterness of the cup of death, and I will never forsake him until he hath arrived in the kingdom of heaven. If evil spirits essay to seize him, I will be his defender on the day of his tribulation.

The man who hath clothed the naked in thy name, I will array in the invisible apparel of life, which will neither fray nor wear out, and I will crown him with an eternal and everlasting crown.

The man who hath given bread to the starving in thy name I will feed on the bread which is not made with human hands.

The man who hath given drink to the thirsty in thy name I will make to drink a cup of the water of life which bubbleth up in the Garden of Eden, and which is sweeter than honey and sugar.

The man who hath comforted the afflicted in thy name will I comfort when he is a sufferer from grief and pain.

The man who hath made the sad to be cheerful through thee I will make to rejoice in my kingdom and in that of my heavenly Father.

I will write in the Book of Life the name of him that shall have caused to be written, or shall himself write, the praises of thee.

I will light in the kingdom of heaven for the man who hath given a lamp [to a church dedicated to thee] a lamp which shall shine seven times brighter than the sun in the kingdom of heaven [and be], like unto the moon [Isaiah xxx. 26].

I will grant My favour before beings celestial and beings terrestrial to the man who shall give thy name to his daughter.

The place wherein this prayer is, or where thy name is invoked, or where the image of thee is set up, or where a commemoration of thee is celebrated, shall not be approached by the powers of evil spirits; and all the filthy hosts of darkness and the spirits that work evil shall flee far therefrom.

The might of the Enemy shall neither attack nor prevail over the man who carrieth this prayer. The evil spirits shall not come nigh unto him, and no foul or filthy spirit, and no spirit of the night or day, whether they make themselves visible by a thrust of a thorn, or by a stamp of the feet; or by a dream by night or by day; or by the impurity (?) of bread, or by the foulness (?) of water or wine; or by drunkenness or wrath, whether it be by sickness or headache, or toothache, or by a foul mouth or pain of the heart; or by small-pox or by disease in the hands and feet; or by deadly fever or by a running cold; or by stomach-ache or by shivering; on sea or on land, among trees or rocks, or fire or water; by arrogance or pleasure, or merry-making or hatred; by the howlings of wild beasts or the cries of the birds; by the heat of the sun or the chill of ice and snow; by blasts of wind, by the bites of dogs and snakes and cobras and scorpions; by the blazing of fire and the flowing of blood; whether it be in the darkness of the night or in the light of day, none of these spirits shall attack the man who carrieth this prayer, and the evil eye shall pass him by. If a man reciteth this prayer thieves of grain shall not come nigh unto his fields to steal wheat or barley or any other crop. Even so shall it be in the case of the wild animals which attack by day or by night, and if they come upon him they shall not be able to harm him; and even so shall it be in the case of hail storms and the attacks of grasshoppers [locusts and such-like], for none of the above-mentioned evils can draw nigh to the man who carrieth this prayer. All those who carry this prayer shall be protected from murrain in his cattle, and drought, and disastrous capture [of beasts?]. And I will save him that carrieth this prayer from every calamity, and every kind of suffering, and he shall escape fatal illness. If he be attacked by a disease that can be cured I will heal him quickly, and if he hath committed sins they shall be forgiven him. If his disease is incurable I will send to him angels of light who shall carry his soul to a place of light and bring it to me; the bad angels shall not go near him, and the spirits of evil that dwell in the Third Heaven shall not lay claim to him. I will lay claim to him and will be his guide on the day of his trouble, and I will go with him to my Father and the Paraclete. And with Me shall come the Twelve Ranks of angels wearing collars of gold, and bearing censers, and wearing rings of gold and rich apparel, and crowns of gold and spikenard in the form of the rainbow, some made of fire and some of lightning, into the Fifth Heaven, to receive the man who carrieth this prayer. I will take him upon my breast, and I will make him to traverse [in safety] the sea of fire, and I will bring him before my throne. When the hosts of heaven see him they shall utter cries of joy, and they will wave their wings and strike the ground with their feet, and rejoice over him that hath carried this prayer.

O, my Mother, didst thou not comprehend what I spake in my Gospel, saying, If a man who hath one hundred sheep loseth one of them, will he not leave the ninety-and-nine in the desert and go and seek the sheep which is lost? And when he findeth the sheep, he taketh it up upon his shoulders, and rejoiceth over it more than over the ninety-and-nine which he hath not lost. Then he calleth his friends and neighbours and saith unto them, Rejoice with me, for I have found the sheep which was lost. Verily I say unto you, there shall be more joy in heaven over one sinner who repenteth than over ninety-and-nine righteous men who do not need repentance (Luke xv. 4–7). All the celestial hosts shall rejoice over him that hath carried this prayer. When his soul shall go forth from his body and he shall depart from this fleeting world, I will bring him to my holy mountain (Isaiah lvi. 7), and I will make him to be acceptable to my Father. Mercy, compassion, grace and everlasting gladness shall be where this prayer is. The places where this prayer is recited shall be free from the plague, and pestilence, and deadly diseases of every kind, no matter what their names may be. I will bless him that carrieth this prayer, and his wife, and his children, and all his possessions, [and I will grant to him] everything which he shall ask in thy name by this prayer and by this writing whether he washeth, or invoketh aid [against evil spirits], or drinketh, or lowereth his voice, or sprinkleth water in his house with a pure heart, and a right faith, doubting nothing. I will consider his prayer forthwith, and I will grant him his heart's desire. MICHAEL and GABRIEL shall go to him and minister unto him wheresoever he may be; and all the hosts of angels shall come and watch over carefully him that shall carry this prayer. O my mother, thou Virgin MARY, who didst give Me birth, I thy partner hereby give thee everything, and I bestow upon thee glory both in the heavens and upon the earth."

And the blessed Virgin asked Him, saying, "Dost Thou say this, O my Son?" And the Lord JESUS replied, "I swear unto thee, and I will not lie unto thee, O MARY, my mother. I swear unto thee by GOD, my Father, by CHRIST, which is my name; by the Paraclete, the Holy Spirit; by MICHAEL, the angel of my wisdom; by GABRIEL, who announced my birth; by the Four Beasts with six wings and many eyes who carry my throne; by the Four-and-twenty sages, who cense my throne and praise the glory of my Being; by the ten thousand angels who stand at my right hand; by the ten thousand who stand at my left hand; by the ten thousand who are before me; by the ten thousand who are behind me; by the tens of thousands of tens of thousands of angels and by the ten thousand who watch; by the first ADAM, my first-born; by ABEL, and SETH, and CAINAN, and MAHALALEEL, and ENOCH, and YARED, and METHUSELAH, and NOAH, with whom I made a covenant in the heavens and on the earth, saying, 'I will never again destroy the earth by the waters of a Flood' (Gen. ix. 11, 15). I swear unto thee by MELCHISEDEK my priest and my type; by ABRAHAM, my beloved; by ISAAC, my servant; by JACOB, my holy one, in whom I planted twelve branches; by JUDAH, by PHAREZ, by BENJAMIN, by LEVI, by ISSACHAR; by the people of the Twelve Tribes of ISRAEL; by all the holy and righteous Fathers; and by ENOCH and by ELIJAH, the writers of my commandments. I swear unto thee by thy pure bowels wherein I dwelt for nine months and five days; by thy breasts whereat I drank milk which was sweeter than honey and sugar, and whiter than the water of Eden. I swear unto thee by the one hundred and forty thousand children of Bethlehem which HEROD had slain for my sake; and by the fifteen prophets who have proclaimed my kingdom; and by my envoys the Twelve Apostles; and by all my disciples who sacrificed their lives for my sake; by the heavens, my dwelling-place, and by the earth whereon my feet rest;

by the nine-and-ninety Ranks of angels; by the flame of fire of my veil; by the heights of heaven which are my habitation; by the shedding of my blood and by the sorrow of my death; by my sojourn for three days in the womb of the earth; by my descent into hell; by my going forth from the tomb; by any resurrection from the dead on the third day; by my ascension into heaven; by my second coming in great glory; by my holy Body; by my holy Blood; by Jerusalem [the city] set free; by Sion decorated with glory; by the Sabbath of the Christians whereon I was born, and baptized, and revealed my Resurrection for life and salvation; by the Holy Church, the bride adorned; by Mount Sion; by the Mount of Olives, the door of my Kingdom; by Golgotha, my tomb; by Mount Tabor, my abode, whereon my Transfiguration took place; by the Christian Church, my Bride; and by thy white and shining form. By all these things I swear unto thee, O Virgin MARY, my mother, who didst bring me into the world, that I will not deceive thee by my promise, that my word to thee shall not prove a lie, and that I will never forget the declarations which I have made unto thee. If a shrine be built and dedicated to thy name, I will dwell therein and will accept the sweet savour of its offerings as I accepted those of ABEL, the righteous man."

The blessed Virgin answered and said unto Him, "Blessed be Thou as are Thy Father and the Holy Spirit, O Thou Who hast granted unto me all these things of Thine own free will. Praise be unto Thee, O Lord, and glory be to Thy Kingdom, and to the life-giving Spirit. Praise be unto Thee, O heavenly Father, now and always and for ever and ever. Amen."

Our Lord, having finished His converse with His mother, gave her the salutation of peace, and went up into heaven with great glory. And the Virgin went back to her house with great joy, and she praised God, saying, " Blessed be Thou, O Lord. May Thy name be blessed and glorified, with Thy Father and Holy Spirit, for ever and ever. Amen."

INDEX

ዘአሀሎ፡እብ፡ወ	አምትሀስተሀ
ወልድ፡ወመንፈ	ገጓል፡ማርህም
ስ፡ትዱስ፡፩አም	እንት፡ታበወዕ
ላክ፡ጸሎት፡በእ	ውስተ፡ጿባዘ
መዩ፡ኃኒት፡መ	አንቀጸ፡ወተብ
ጴሐፈ፡ሒሀው	ጴሕ፡ወስተ፡መ
ት፡ዘትሰመዩ፡	ንጓሠተ፡ሰማ
ልፋፈ፡ጼዩቅ	ያት፡መርሒ፡ሳጸ
ዘጼሐፈ፡አብ፡	ዩት፡ወዘንተነ
በእየዊሀ፡አም	ገፈ፡ስማርያም
ትሀመ፡ሒትወ	እወ፡እምሀ፡ባ
ላሀ፡ክርስቶስ	ረ፡ተወልሀ፡እ

Lefâfa Ṣedeḳ (A). Brit. Mus., MS. Add. 16,204, Folio 1*b*.

ምህኒሃ ፨ አመ፡ ፲ወ
፫፡ ለየካቲት፡ አ
ስተርአየ፡ ክር
ስቶስ፡ ለማርያ
ም፡ ነበ፡ ደነብሩ
ጺዮቶን፡ ውስ
ተ፡ ገነት፡ ወነ
በ፡ ደነብሩ፡ ኗዋ
አን፡ ውስተ፡ ዲዩ
ን ፨ ወትቤ፡ እግዚ
እትነ፡ ማርያም፡
ወሶበ፡ ትሬኢ፡ ዲ
ንገፀት፡ ወርዕደ
ት፡ ወፈርሃት፡
ዓበየ፡ ፍርሃት፡
ወደቤላ፡ ኢየሱ

ስ፡ ለማርያም፡ አት
ፍርሂ፡ ኦማርያም
እምየ፡ እንተ፡ ዐ
ርክነ፡ በክርሥ
ኪ፡ ሠመለዩ፡ ክነ
በመንፈስ፡ ቅዱ
ስ ፨ ወትቤሎ፡ በ
እንተ፡ ምንት፡ ዐ
ርኩኒ፡ ገጥሬኒ
ኦወልድየ፡ በም
ንት፡ ይደኅኑ፡ አ
ዝማድየ፡ እምዐ
ዝነኔ፡ እሳት፡ በ
ለዓ ፨ አንስ፡ አፈ
ርህ፡ በእንተ፡ ነፍ
ስየ ፨ ወበእንተ፡

LEFÂFA ṢEDEḲ (A). Brit. Mus., MS. Add. 16,204, Folio 2a.

ግኒቶ፡ፈፈ፡ጥጄ መ፡ ጴያ፨
ⵁⵁⵁ መ

ኢ፡ያቲ፡ምህ፡ እቡ፡ የ፡ ጎረ፡ከ፡፡ እስመ፡
ወበእንተ፡ ሒሩ፡እ ዘተናገሩ፡፡ ሂ፡የ፡
ም፡የ፨ወበእንተ፡ ወጸ፡እ፡ኀበ፡ ሣል
ሳሙ፡ኢ፡ል፡ወየ፡ ሀየ፨ወ እም ህ፡
ሲፍ፡እ ኃው፡የ፨ ኀረሁ፡ስ፡ የ፡ዘረ፡
ወበእንተ፡ እ፡ልሳ ዐ፡ው፡ስተ፡ኵሎ፡
ቢጥ፡ እኣት የ፨ ወ ሰብእ፡ወ የ፡ገብ
በእንተ፡ ዳዊት፡ ጋ ሩ፡ ኃጠ፡ እተ፡እ
ገጸ፡መ፡ላህ፡የ፨ ነዘ፡የ፡ ብሉ፡ ሀለ
ወየ፡እዚ፡ነ፡ነጎ ወነ፡ በዘንህ፡ ባን
ረነ፡፡ ወልድ፡የ፡ጥ ቦቱ፨ ወካዕበ
ሁ፡ቱ፡ በዘሁ፡ዱ፡ባ ተ፡ስእለት፡፡ ወት
ሉ፡ እም፡ ዘንቱ፡ እ ቢሎ፡፡ ማርሃም
ሳት፡በ ላዊ፨ ወሀ፡ በእንተ፡ ያር፡ከ፡ከ
ቢስ፡ ኢ፡ሀሁ፡ስ፡ለ በክር፡ሠ፡የ፡ሀ እ
ማር፡ ያም ፡ እ፡ ሂ፡ነ ው፡ረ፡ ኍ፡ ወሪዕለ

ተ፨ወበከየት፡እ
ግዝእትነ፡ማር
ያም፡አንብባ፡መ
ሪረ፡ወክርስቶ
ስኒ፡በከየ፡ምስ
ሌሃ፨ወይቤላ፡
ኢትብክዪ፡ማ
ርያም፡እምየ፡ና
ሁ፡እነግር፡ለአ
ቡሃ፨ወለእመሰ
አብሁኒ፡እነግሬ
ከ፡ወሖረ፡ኀበ፡
አቡሁ፡ወይቤሎ
ናሁ፡ማርያም፡እ
ምየ፡ትበኪ፡ሀበ
ኒ፡መጽሐፈ፡ሕ

ይወት፡እኅት፡ጽ
ሐፋከ፡በእዴከ፡
ቅድስት፡እምቅ
ድመ፡ትትወለድ፡
አነ፡እማርያም፡ድ
ንግል፡ትንብር፡ዲ
በ፡ሥረገላ፡ኪሩ
ቤል፡መንበርከ፡
ወይቤሎ፡አቡሁ፡
ለወልዱ፨ናሁ፡
ወሀብኩከ፡ሑር
ንግራ፡ለማርያ
ም፡እምከ፡አልቦ
ዘኀብእኩከ፡እም
ኔሃ፨መፍትው፡
ለከ፡አሳ፡ከሠት

LEFÂFA ṢEDEḲ (A). Brit. Mus., MS. Add. 16,204, Folio 3*a*.

ኩ፡ለከ፡ኵሎሙ፨
ወጸሐፊ፡ኢየሱ
ስ፡ክርስቶስ፡በቀ
ለመ፡ወርቅ፡ወ
መጽአ፡ዩመና፡ብ
ሩሀ፡ወጸለሎሙ
ወግብሩ፡ኀ፡መን
ጦላዕት፡ዘእሳት፡
ወአልቦ፡ዘአእመ
ሮ፡ወኢሰምዑ፡
ኢመላእክት፡ወ
ኢሊቃነ፡መላእክ
ት፡እስከ፡ይነግራ
ዘንቱ፡ነገረ፡ለሐዋ
ርያሁ፡ወይቤላ፡ን
ሥኢ፡ዘንቱ፡ዘወ

ሀብኩኪ፨ወአኙ
ቴኒ፡ኢትክሥቴ፡
ለዘኢይክል፡ፀዊ
ሮቶ፡ወዓቂቦቶ፡
ለዛቲ፡መጽሐፍ
እንጋአ፡ለጠቢባ
ን፡አለ፡የአምኑ፡
ብየ፡ወእለ፡የሐ
ውሩ፡በትእዛዝ
የ፨ወዘአጥረያ፡
ለዛቲ፡መጽሐፍ
ኢያወርድ፡ው
ስተ፡ደይን፡ወኢ
ውስተ፡ሲኦል፨
እመሂ፡ዘፃራ፡ወ
ዘዓነቀ፡በክሣዱ፡

LEFÂFA ṢEDEḲ (A). Brit. Mus., MS. Add. 16,204, Folio 3b.

ይት፡ኃሀግ፡ሉቱ ት፡ቀወየበጽ፡ሐም
ኃጠ፡አቱ፡ቀወለ ቅድመ፡አንዚአ
እመ፡ደገመ፡በ ብሔር፡ወየበው
ታሉ፡በጊዜ፡ተነ ዕም፡ውስተ፡መ
ርባን፡ይት፡ኃየ ንግሥተ፡ሰማያ
ኃ፡ወየ፡ነጽሐ፡ ት፡አብዓሊ፡ሲተ
እምርስሐተ፡ኃ ላ ገ፡ብርክ፡እዕጠ
ጢ፡እተ፡ቀወለአ ሉ ዕለ ው
መ፡ግብሩ፡ንበ አተ፡ብርሃነ፡ሐየ
ወኃንዙ፡ማዕተ ወት፡ወመይ፡ኃኒ
በ፡ሰሉ፡ሞን፡፫ ት፡ዘለሃለም፡ቀወ
በዛቲ፡መጽሐፍ ዘንተ፡ፈጸሞ፡አ
ለእመ፡ተቀብረ የሱስ፡ነገራ፡አስ
ደመርሐው፡መ ማቲሀ፡ዘድልው
ላእክት፡ውስተ ለሐየወት፡ወለ
አንተጽ፡ሐየወ መድኃኒት፡ቀወ

LEFÂFA ṢEDEḲ (A). Brit. Mus., MS. Add. 16,204, Folio 4a.

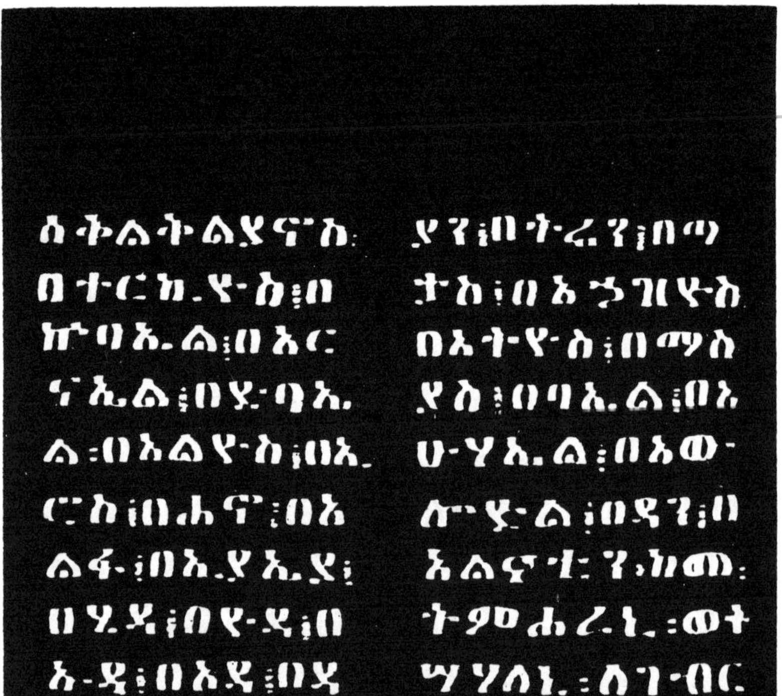

ሰቅሎቅሎያኖስ፡
በተርካዩስ፡በ
ኰባኤል፡በአር
ናኤል፡በጁባኤ
ል፡በአልዩስ፡በኤ
ርስ፡በሐና፡በአ
ልፋ፡በኤያኤያ፡
በሂዲ፡በዩዲ፡በ
ኡዲ፡በአዲ፡በዲ
ልዴ፡በሐሪ፡በዴ
ኔ፡ሰወላዲ፡በኒ
ባ፡በአልፋ፡ወኔዮ
ዶሐሪ፡በዴልዲ፡
በአኡሀዲዴልያ
ዴ፡በ፤ዮልኬን፡
በሀሀዲዲ፡በአው

ያን፡በትራን፡በጣ
ቱስ፡በአጎገዮስ
በኤቶዮስ፡በማስ
ያስ፡በባኤል፡በኤ
ሁሃኤል፡በአው
ሎሂል፡በዳን፡በ
አልፍቲ፡ንክመ፡
ትምሕረኒ፡ወት
ሣሃለኒ፡ለገብር
ከ፡እስጢፋኖስ
ወደቤሳ፡ኢየሱ
ስ፡ክርስቶስ፡ስ
ሀሃርያ ምን ዳጹም
ምሕረት፡ለአበ
የ፡ሰማያዊ ወ
ለእሙነ፡ተአጠ

LEFÂFA ŞEDEĶ (A). Brit. Mus., MS. Add. 16,204, Folio 5*a*.

Lefâfa Şedeķ (A). Brit. Mus., MS. Add. 16,204, Folio 5b.

በኄ፡ በዋጠ፡ በዛ
ሆ፡ በሐ፡ት፡ በጤ
ት፡ በየዩ፡ በኑ
ፍ፡ በሰሚሂ፡ በ
ሚያሙ፡ በኖኘ፡ በ
ሳያኬት፡ በዔ፡
ዘሬ፡ በጴዲ፡ በቶ
ፍ፡ በሬስ፡ በሳኘ፡
በታጡ፡ ለትቅሬ
ብ፡ ስአለትዩ፡ ነ
ቢከ፡ እግዚኦ፡
በኳዴለ፡ ዝኘቴ
አስማቲከ፡ እኁት
ር እሄኒ፡ ጠሶ፡ ለ
ዴዴኘ፡ ለገብርከ፡
ኦስጠ [illegible] ስ
ኦስማቲከ፡ እግዚ
አብሔር፡ ዓበዴ
ዝወእቱ፡ ዘለዓለ
ም፡ እግዚእነ፡ ዘ
ነገሮ፡ ለጴጥሮስ
ሰማያዊ፡ አፎስ፡
በፍኪዴሮ፡ በሳ
ሁ፡ በያሁስዷዶስ፡
በአፉኘ፡ በአፍሳ፡
በአሰና፡ በአፉሳ
ቁኘ፡ በሳህሳሁ፡
በሳእናሔነጮ፡ በ
ኘሕልፍ፡ በአርዮ
ስ፡ በወርዮስ፡ በ
አክልያ፡ በጴልያ፡
በተሣሃሎ፡ ሚ

LEFÂFA ṢEDEḲ (A). Brit. Mus., MS. Add. 16,204, Folio 6a.

ዶ፡ በሔኢ፡ በአዮ፡
በርሚኽርሚር፡
በሱርያል፡ በሰዲ
ቋኤል፡ በሰሳትያ
ል፡ በኦፋኽያል፡ በ
አንያል፡ በሚልማ
ኤል፡ በአጥዮዶ፡
ኢዩ፡ ልሳን፡ በኤል
ፋጢ፡ በአኢ፡ ዶኁ
ራጢ፡ ስዕክ፡ በያ
ው፡ በኤጓዮስ፡ በ
ካፊ፡ በአርዕኝ
ያል፡ በስዕያል፡
በአፋፉ፡ በአራና
ቱ፡ በአፋራስኮር
ስ፡ በአሂ፡ በኤሎ

ሂ፡ በአፋኒያል፡
በኤማኑኤል፡ በአ
ብርስትያል፡ በአ
ልያል፡ በኤርናኤ
ል፡ በአማስርዮል
በአፋሳርዮል፡ በ
ጓርምዩል፡ በጸር
ምልዩል፡ በቶርጸ
ልዩል፡ በጓርመ
ልዩል፡ በጸርአሰ
ጢስ፡ በአርክያል፡
በሰርስያስል፡ በአ
ንዮስ፡ በጤቤር
ዩ፡ በሂትዮ፡ በጥ
ርስጸምን፡ በማር
ያ፡ በማርማ፡ በአ

Lefâfa Ṣedeḳ (A). Brit. Mus., MS. Add. 16,204, Folio 6b.

ንሶስ፡በዳኪ፡በአ
ቡደቴር፡በሔራ
ጦን፡በጰንከተራ
ጥር፡በኢያስዩን
ሮድኑ፡በኑዱራ፡
በኡኡሱሱናያ
ክአእዩዎስ፡በሰ
ለስኤል፡በሂሲዎ
ን፡በዲንፉስ፡በኃ
ደለ፡ዝንቱ፡አስ
ማቲከ፡ተማኅ
ፀንኩ፡አነ፡ገብ
ርከ፡ወልደ፡ሚ
ካኤል፡ዮር፡አለ
ዮር፡ዲናት፡አዲ
ራ፡ሮዳስ፡በ፳

ቅንዋተ፡መስቀ
ሉ፡ለእግዚእነ፡
ኢየሱስ፡ክርስ
ቶስ፡ተማኅፀን
ኩ፡አነ፡ገብርከ
አሰ

በስመ፡አብ፡ወወ
ልድ፡ወመንፈስ፡ቅ
ዱስ፡፩አምላክ፡

ጸሎት፡በእንተ፡
ባዕረ፡ሞት፡ዲዮ
ስ፡በትሮን፡ኩፉ
ሃ፡ጋኖን፡ከጡስ፡
ቴርሎ፡ወአዲል
ክፋፆ፡ለበዴን፡
ለዘቲ፡መጽሐ
ፍ፡ዘያነብብዋ፡
በደኃሪት፡ዕለ
ት፨አመ፡ዕለተ፡
ኵነኔ፡ጉባኤ፡በሃ
ጎግ፡እለ፡ያስሕ
ቱ፡ሕገ፡እግዚ
አብሔር፨ወእለ
ያመጽኡ፡ነገረ፡
ጠዋየ፨ወደብ
ሉ፡አነ፡ውእቱ፡
ክርስቶስ፡ወል
ደ፡እግዚአብሔ
ር፡ሕያው፨ወየ
አምንዎ፡ኵሉ
ሙ፡ኖሎኒ፡ሕ
ዝበ፡ክርስቲያን
ደብሉ፡መንሕነ
ስ ነአምን፡በስ
መ፡ኢየሱስ፡ክር
ስቶስ፡በጠልዩ፡
እግዚአብሔር፡
በአብ፡ወበወል
ዱ፡ወበመንፈስ
ቅዱስ፨ወኢል
ያስ፡ያሰብክ፡ለ

LEFÂFA ṢEDEḲ (A). Brit. Mus., MS. Add. 16,204; Folio 7b.

ኩሎ ፡ ሕዝብ ፡ ክር
ስት ፡ ደገ ፡ ወሃ አም
ነም ፡ ስክርስቱስ
ወልደ ፡ ወበወል
ደ ፡ ሰዶጣንሶ ፡ ዘ
የአምነ ፡ ዲ ት ፡ ነ
ነ ፡ በደዬገ ፡ ወዘ
ሰ ፡ የአምነ ፡ በእ ፡
የሱስ ፡ ክርስቱስ
ወልደ ፡ እግዚአብ
ሔር ፡ ኢየበውዕ
ውስተ ፡ ሃዮን ፡
ሃደልም ፡ ወሃሔ
ውር ፡ በመገራ ፡ሕ
ትዲስ ፡ የበ ፡ እግ
ዘ አብሔር ፡ አለ ፡

ወ እተ ፡ አያሰከ
ሰማዬ ፡ ወያሂ ር
ወአንሶሰወ ፡ ኅት
ራኤል ፡ ገ ገ ሥ ፡ ካ
ሃሁ ፡ ዲ በኪ ፡ ክር
ስት ፡ ደናዊ ፡ ክለማደ
ዲ ፡ አዘት ት ፡ ክ ብር
ወ ሕደ ወት ፡ ዝከ
ዘ ደዋቢ ገ ፡ አና ራ
ሰ ፡ ሕደወ ት ፡ አመ
ዕለት ፡ ዓ ዳ ፡ ወደሃ
ነ ፡ ወበወ እት ፡
መዋዕል ፡ በሔደ ፡ ነ
ዲ ፡ ጸልም ፡ ወወ
ረ ነ ፡ ሃሜደ ፡ ሃ ክወ
ነ ፡ ወበወ እት ፡

Lefâfa Ṣedeḳ (A). Brit. Mus., MS. Add. 16,204, Folio 8a.

Lefâfa Ṣedeḳ (A). Brit. Mus., MS. Add. 16,204, Folio 8*b*.

·ቆኡቡሃ፡ኄልቀ
ነፍሶሙ፡ለእለ
ያስተኀቅሩ፡ቃ
ሎ። ወይቤሎ
ሙ፡ለአበዊነ፡
አመ፡ዕለተ፡ፋ
ዳ፡ወደይን። ብ
ፁዕ፡ውእቱ፡ዘ
አጽሐፎ። ብፁ
ዕ፡ውእቱ፡ዘጣ
ነቆ፡በክሣዱ፡
ወዘተአመኖ፡ለ
ዝንቱ፡መጽሐ
ፍ፡ኢይለክፎ፡
ገሃነም። ወበሙ
እቱ፡መዋዕል፡

ፀሐይ፡ዘኢየዓ
ርብ፡ወማኅቶ
ት፡ዘኢይጠፍዕ
ቃለ፡ፍዳሆሙ፡
ዘኢየረምም።
ስብሐት፡መንግ
ሥቱ፡ዘኢይት
ነሰት። መንበሩ፡
ዘእሳት፡ክሉል
ዕ ዘኢያንቀለቅ
ል፡ለዓለመ፡ዓለ
ም፡አሜን። ወ
ይቤሎሙ፡መላእ
ክቲሁ፡ዘንዜኑ
ስመነ፡ከመ፡ንሴ
ብሖ፡ወንዜም

ር፡ለክ፨ወዲቤ
ሎሙ፤ቀዲሳሞዊ
ስምየ፡አዶያዋዲ፡
ካልዕ፡ስምየ፡ኪ
ገዶ፡ሣልስ፡ስም
የ፡አማኑኤልፉ
ብዕ፡ስምየ፡ኢዮ
ሱ፡ስ፡ኃምስ፡ስ
ምየ፡ክርስቶስ፡
ሳድ፡ስ፡ስምየ፡ሶ
የዴ፤ሳብዕ፡ስም
የ፡እግዚአብሔር፡
ለእሙ፡በዝንቱ፡
አስማቲ፡ዘትአ
መኑ፤ወዘገብረ
ተዝካርየ፨እም

ሔር፡አነ፡እምዋ
ንቱ፡እሰትነ፡ንዲዲ
ወዕፁሁ፡ዘኢዮ
ነጉም፡ወእሳት፡
ዘኢዮጠፋዕ፡
ወበሱ፡ዘኢዮ
ደክም፨ወዲበ
ሎ፡እግዚአብሔ
ር፡ለማኅሌል፡
ወሀብኩ፡ኪ፡ማ
ዕኩቶየ፨ወለ
እሙ፡ዘገብረ
ተዝካርየ፤ወዘ
ተአመነ፡ኪዳንየ፡
ወለዝንቱ፡መጽ
ሐፍ፡ዘሢኖቹ፡ወ

Lefâfa Ṣedeḳ (A). Brit. Mus., MS. Add. 16,204, Folio 9b.

Lefâfa Ṣedeḳ (A). Brit. Mus., MS. Add. 16,204, Folio 10a.

ተሰማዪ፡ ወዘአ
ኀበለ፡ ፴ወ፫ ጊ ሳው
ያን፨ ወራቶሔምዋ
ኀተማዓሃ፡ ወነጸር
ዋ፡ ወአንበበዋ፡
ከመ፡ ዩስያሙ
ወሰቢሃ፡ ነሥአ
ጤሳእክት፡ ፫ቱ፡
መጣቅዕት፡ ወ
ጠቅሁ፨ ወነሥ
አ፡ ፫ቱ፡ ጸዋዓተ
ወከዓዉ፡ ወስ
ተ፡ ገጸ፡ ምድር፡
ከመ፡ ዩትቀኗሱ
ወለዩ፡ ኒሬን፡
ወዩቶሬ ለጠ ሰ

ማዩ፡ ወያዘጁር፡
፫ቲ፡ ማዕፀቲ፡
ወ፫ቲ፡ በርሃና
ት፡ ወ፫፡ ያውሰዋ
ሬ፡ መዝበሩ፡ ለኤ
ግዚአብሔር፡ በ
ዘተእምኑ፡ ስሞ
ግሩሙ፡ በነበ፡ ሀ
ለዉ፡ ነበዪት፡ ወ
ሔዋርያት፡ ወዩ
በር፡ ልዑል፨ አ
መ፡ ፲ወ፯ ለመስ
ከረያ፡ ቀዲሱ፡
ሥጋሃ፡ በንጹሕ
በእንተ፡ መስቀ
ሉ፡ ለክርስቶስ፡

ክቡር፨ወለመቃ
ብረ፡ እግዚእነ፡
ኢየሱስ፡ ክርስ
ቶስ፨ያስተርኢ
ምሕረተ፡ ለዕሌ
ነ፨በከመ፡ ቀል
ከ፡ ቅዱስ፨ወዮ
ሐ ለሙ፡ ለቅዱ
ሳኒሁ፡ በሉሬዎ
ከበራት፡ ቃለ፡
ለእግዚአብሔ
ር፡ አማኖራ፡ ዝ
ምራኤል፡ ጋር
ካኤል፡ ዱምና
ኤል፡ ኪዳ፡ አሄ
ናኤል፡ ኒሩት፡

ዝብድዮስ፡ ሱ
ሞኚዮስ፡ ማዲል
ተራ፡ ታርቦታኪ
መያትር፡ ጓፍያ
ኖስ፡ አርጵራ፡ ኀፍ
ያሂ፡ ቀታዊር፡
ወርያኤል፡ ኦል
ዳኚ፡ ስመ፡ አተ
ዋስ፡ ሰሰሮ፨ወ
ከመዝ፡ ፍክሬሁ፡
በግዕዝ፡፡ ወበእ
ንተዝ፡ ዕርጋት፡
ለሠዋርያም፡ ከማ
ሁ፡ አዕርገነ፡ ለ
ተ፡ ለገብርከ፡ ወ
ሊዩ ሠዋ ካሌል፡

LEFÂFA ṢEDEḲ (A). Brit. Mus., MS. Add. 16,204, Folio 11a.

፩ ስሙ ፡ ሰፍኖ
ስ ፡ ፩ ስሙ ፡ ቆሆ
ኬ ፡ ወ፩ ስሙ ፡ ገ
ብርኤል ፡ ወ፩ ስ
ሙ ፡ ብርሃናኤ
ል ፡ ወ፩ ስሙ ፡ ጹ
ራኤል ፡ ወ፩ ስሙ
ዝምራዳኤል ፡
ወ፩ ስሙ ፡ ደዶ
ያ ፡ ዜናቲ ፡ አስማ
ት ፡ አሀሉ ፡ ውስ
ተ ፡ ልበ ፡ ሰብእ ፡
መዋትያን ፡ ዛቲ ፡
ደአቲ ፡ ዘወጹአ
ት ፡ እምአፉሆሙ
ወእምቃለ ፡ ክና

ፋሬሆሙ ፡ ለአብ
ወወልድ ፡ ወመን
ፈስ ፡ ቅዱስ ፡ ፩ ስ
ሙ ፡ አግዮስ ፡ ወ
፩ ስሙ ፡ አርህና
ን ፡ ወ፩ ስሙ ፡ በ
ትሮን ፡ ወ፩ ስሙ
አስራሮን ፡ ወ፩ ስ
ሙ ፡ ጾኑ ፡ ወ፩
ስሙ ፡ ምክያር ፡
ወ፩ ስሙ ፡ ምዩ
ዮስ ፡ ወ፩ ስሙ ፡
አግዮስ ፡ ወ፩ ስሙ
መዋትልሌም ፡
ወ፩ ስሙ ፡ እል
መክን ፡ ወ፩ ስሙ

LEFÂFA ṢEDEḲ (A). Brit. Mus., MS. Add. 16,204, Folio 11b.

እያ፡ወ፩ስሙ፡
ምክያሮ፡ወ፩ስ
ሙ፡ጋኖን፡ወ
፩ስሙ፡ነጻዲሃ
ለነስ፡ወ፩ስሙ
አዳሂል፡ወ፩ስ
ሙ፡ማምአያዩ
ስ፡ወ፩ስሙ፡አ
ነትዩር፡ወ፩ስ
ሙ፡ኪዲዮሮ
ስ፨ዛቃቢሃ፡ለ
ነፍስ፡ወእናቅጼ
ሃ፡ዘንቱ፡አስማ
ተ፡ዘዴፀው፡ሬ፡
በየውሃት፡ወበ
አርምሞ፡ወበ

ትዕግሥት፡ወ
በለኖሳስ፨መበ
ሬሬሃ፡እማዚአ
ብሔር፡ያዴኅን
ዘንቱ፡ቃለ፡ሰማ
ያ፡በእሴ፡በእዘ
ኒሁ፡ደማየኘ፡
በወርቅ፡ወበብ
ሩር፡ወበአልባስ
ክበራት፨ወዘ
ንተ፡ለእመ፡ኳፕ
አ፡ደኩን፡ገብረ
ሙ፡ልስቱ፡ደያው፡
ስምየ፡ህዋ፡ዘአ
ጥመቀ፡አባ፡ያሔ
ንስ፡በዛቲ፡ዕለት

LEFÂFA ṢEDEḲ (A). Brit. Mus., MS. Add. 16,204, Folio 12*a*.

ወበዛቲ ፡ ሰዓት ፡
ደራቲሒ ፡ አንቀጸ
ጸጹቀ ፨ ወኢደ
ረእደዎ ፡ ሰደደን
ወበውስቴ ደደ
ን ፡ ምዕራባሩ ፡ ወደ
ምሕሮ ፡ እግዚአ
ብሔር ፡ ወበእንተ
ዝንቱ ዕርገት ፡
ለመዋርጸያሙ ፡ ከመ
ሁ ፡ ኦዕርግነ ፡ ሰ
ተ ፡ በዝንቱ ፡ አስ
ማቲከ ፡ ተማኅፀ
ንኩ ፡ አነ ፡ ገብርከ
እስጢፋኖስ ፡
ዝሉ አሌል ፡ ስድ

ብኢኤል ፡ ኢዮኤ
ል ፡ ፋናኤል ፡ አቅ
ናኤል ፡ ኢዮብዩ ፡
ቴ ፡ ርሎ ሳኤል ፡ ኢ
ሊሄል ፡ ሰሳቲያኤ
ል ፡ ዕዝራኤል ፡ ቃ
ሳቲሳኤል ፡ አዝ
ራሞ ፡ ኢሳዋ ፡ ኤ
ሳኢራብሳኤል ፡
ስዩራኤል ፡ ሰንበ
ኤል ፨ ብዝንቱ ፡ አ
ስማቲከ ፡ ከመ አ
ደምጸ አነ ፡ ሞት
ወኢሕማም ፡ በዩ
ሳኤል ፡ በልክኤ
ል ፡ በፋሳኤል ፡ በ

Lefâfa Ṣedeḳ (A). Brit. Mus., MS. Add. 16,204, Folio **12*b*.**

ኤኽኤል፡በዴሳ
ፉኤል፡በኤደኤ
ል፡በዶርስሳኤል፡
በዝንቱ፡ኵሎ፡
አስማቲከ፡ተማ
ኅፀንኩ፡ሕነ፡ገብ
ርከ፡እስ[illegible]ሱ
ኢሌሰኤልኮስ፡
ጵንታኮርጢስ፡
አግሚሙስ፡ወ
ጥንትን፡አዕዮን፡
ወአቅማቲስ፡ኢ
ያንኤል፡አዛኤል
ሔኆማ፡መርሞ
ቶናጊ፡አዲራጽ
ብዮን፡በዝንቱ፡

አስማቲከ፡ወበክ
ዕወተ፡ደሙ፡ሰ
ጊዮርጊስ፡ገብር
ከ፡ተዘከረኒ፡እግ
ዚእ፡በመንግሥ
ትከ፡ለገብርከ፡
[illegible]
ሳዶር፡አላዶር፡ዳ
ናት፡አዴራ፡ሮዳ
ስ፡በ፭ቅንዋተ፡
መስቀሉ፡ለእግ
ዚእነ፡ኢየሱስ፡
ክርስቶስ፡ለዓለ
መ፡ዓለም፡አሜ
ን፡

LEFÂFA ṢEDEḲ (A). Brit. Mus., MS. Add. 16,204, Folio 13a.

በአሙ፡አብ፡ወወልድ   ማርያዎ፡በየነሬ
ሀ፡ወበሰ፡ኁሪ፡ስ፡ቅዱ   ት፡ዕላተ፡ወዘንተ፡ኢ
ስ፡ዕእምላክ፡ስ   ስማተ፡ተዘወር
ሞዐ፡ፃነዓሪክሙ   ውስተ፡ክርሡ
አኃዊነ፡ስእሜተ   በእምሳስ፡ጌራ፡
እምነ፡ቀለ፡ልፋ   ወዘዖሪ፡በእምሳ
ፌ፡ጽዴ፡ት፡ዘወ   ሰ፡ማር፡ያህወዘ
ሀባ፡እግዚ፡እብ   ሂ፡ዘባነታ፡ልዛ
ሔር፡ኪዴነ፡ሰ   ቲ፡መጽሐፍ፡ኤ፡

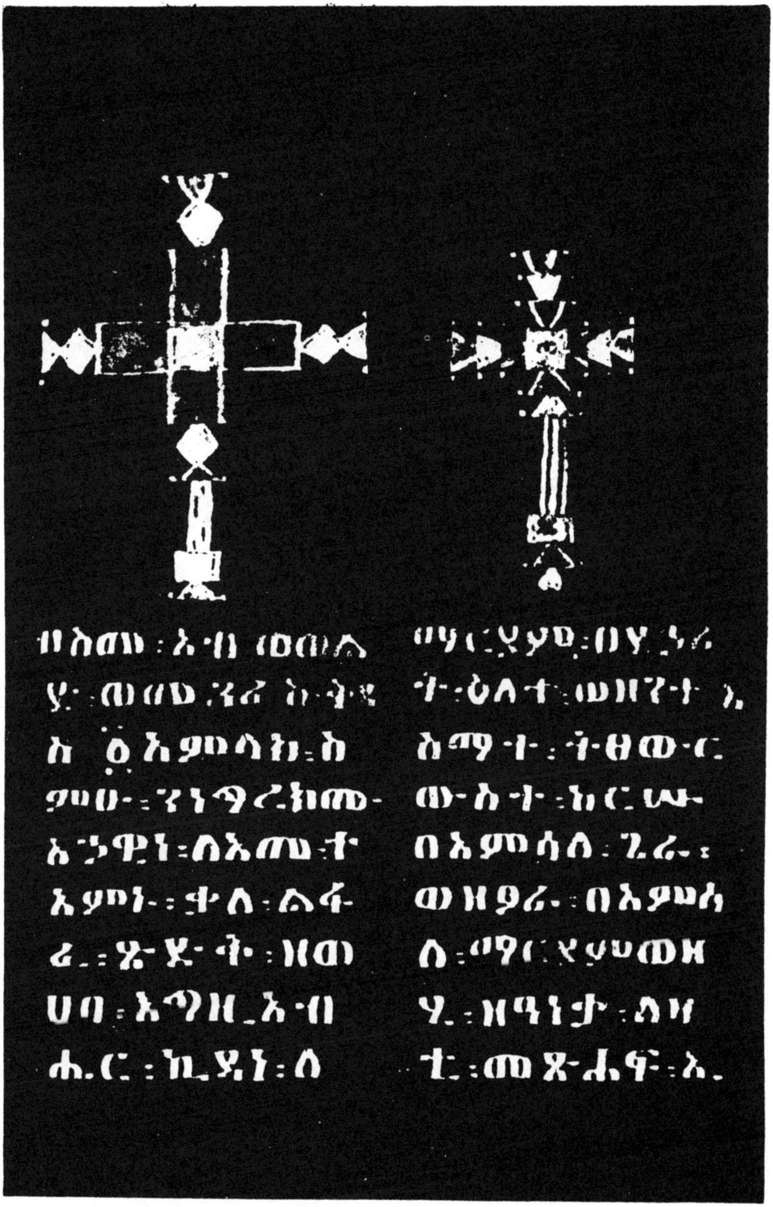

Lefâfa Ṣedeḳ (A). Brit. Mus., MS. Add. 16,204, Folio 13*b*.

ዶሬእዩ፡ለዶዶን
አሳ፡ዶሬክብ፡ሔ
ዶወተ፡ዘለጓለ
ም፨ወተስእለቶ
አግዝእትነ፡ማር
ያም፡ለእግዚእነ፡
ወትቤሎ፡ንግረ
ኒ፡ዘየዓቢ፡እም
ኵሉ፡አስማቲ
ከ፨ወተሰጠዋ፡
እግዚእነ፡ኢየሱ
ስ፡ክርስቶስ፡ለ
ማርያም፡ወይቤ
ላ፡እነግረኪ፡ዘን
ተ፡አስማትየ፡ዘ
ዕፁብ፡ለሰሚዕ

ወኅቡእ፡ለሬእዩ
ወሠናይ፡ለውእ
ቱ፡ለዘይክል፡ፀዋ
ሮቶ፡ወዓቀቦቶ፤
ወነዐበ፡ትቤሎ፡
እግዝእትነ፡ማር
ያም፡ኢስእለከ፡አ
ወልድየ፡ከመ፡ት
ንግረኒ፡ስመከ፡
ኅቡአ፨ወይቤላ፡
እነግረኪ፡ስምየ
ጥዩቀ፨ወአንቲ
ኢታስትቲ፡ዘንተ
አስማትየ፡ዘዕፁ
ብ፡ለዘኢያእም
ን፡በጥሩዕ፡ወለዘ

LEFÂFA ṢEDEḲ (A). Brit. Mus., MS. Add. 16,204, Folio 14*a*.

ኢይዐውር፡ዘገ
ተ፡ነገርየ፡ኢይዴሎ
ዎ፡ይክሥት፡አስ
ማትየ፨ወክዕበ
ተ፡ስእለቶ፡እግዝ
እትነ፡ማሃርያምሰ
ኢየሱስ፡ኢጸነ
ግርማሁ፡ለአብጸ
ን፡ሰብእ፡ለእለ፡
ኢይሌብዉ፡በሎ
ቦሙ፡ወለእለ፡ኢ
የኃሥሡ፡ማኅ
ደረ፡ዘበሰማያት
ወለእለ፡ኢይሜ
ንኑ፡ዘበምድር
ክብረ፤ተሰጠዋ

እግዚእነ፡ለማር
ያምእት፡ፍሎጠ
ይቤሳ፡ከመ፡ያእ
ምሩ፡ዘንተ፡ስም
የ፡ዘእነግረኪ፨
ወሬዙ፡ዓም፡ኢየ
ሱስ፡ቅጠ፡ማዕ
ዝለ፡ዲያምዪ፡ዪመ
ኖ፨ወአስተርአ
ያ፡በነዪ፡ኢሳት፡አ
ስከ፡ዴነግሬ፡ዘ
ንተ፡አስማተ፨
ወዪቢሳ፡ኢሎሂ
ኢሎሂ፡ኢሎሂ፤
ኢሬን፡ኢሬን፡ኢ
ራን፤ሬፌን፡ሬፌ

LEFÂFA ṢEDEḲ (A). Brit. Mus., MS. Add. 16,204, Folio 14*b*.

ን፡ሬ·ሬ፡ን፤ወግዝን
ቱ፡ብሂል፤አንዚ
ኣለም፤ቀሳሪ፡ወ
ዐሐሬ፤ብሂል፤
መርዮን፡ብሂል
አዶትመዓዕ፡ብ
ሂል፡ፎፎሬን፡
ብሂል፤ተሣሃለ
ኒ፤ብሂል፡ብዮ
ን፤ብሂል፤ኒር፤
ብሂል፤በርስባሂ
ል፡ብሂል፨ኵሉ
ዘይሬርሆ፡ስሙ
ለአብ፡ማርያል፤
ስሙ፡ለወልዱ፤
ምናቴር፤ወስሙ

ለመንፈስ፡ቅዱ
ስ፡አብዴቴር፨በ
ዝንቱ፡አስማቲ
ከ፡ተማኅፀንኩ፡
አነ፡ገብርከ፡አሰ
ሩ፡ሮስ፡ ማዕከኤ
ል፤ወገብርኤል፤
ሱራሬል፤ወኪ
ሩቤል፤ሱርያል፡
ወፋኑኤል፤አጸ
ኤል፤ወሰቁኤል
ጎሎቀን፡መለእክ
ት፡ሰአሉ፡ለነ፡አስ
ተምህሩ፡በእንቲ
አነ፨ሰዱቃኤል፤
ብርናኤል፤ሰአሉ

Lefâfa Ṣedeḳ (A). Brit. Mus., MS. Add. 16,204, Folio 15a.

በእንቲአነ፡በጸሎ
ትክሙ፡ክጠ፡ዓ
ደኀኀ፤እግረ፡ማ
ጠ፤ሱርትዩን፤
ሀበሩ ማግረ፤ጠለ
ጠን፡ቦ እንስሳ፡ተ
ማኀፀንኩ፡በአስ
ማቲክሙ፨በአ
ልዳን፡ጠንበርክ
ወበልዎሔሳ፡ሀ
ገርክ፡ወበአርያ
ዎሀ፡ፋ፡ጸ ዎጠ፡ማ
ኅደርክ፨ወበማ
ርያዎበወሳዲትክ
ወበዐወንጊላው
ያን፡በኒወረኔነበያ

ት፡በኒወዩሐዋ
ርያት፡በአወቦክ
ህናተ፡ሰማያ፡ወ
በሠሐረ፡ሰማያ
በርቀወዩ አርድዕ
ት፡ወበረዲበጸ
በርያ፤ወ ተርተ
ዓነ ሃይማኖት
ወበኀለ ቀነ፡ጠ
ለአክት፡ወበየየ
ኦዕላፋት፡ወት
ዕልፈተ፡አዕላፋ
ት፡ወበአስማተ
ሆሙ፡ቅዱሳ
ን፡መላእክት፡ተ
ማኅፀንኩ፡አነ፡

Lefâfa Ṣedeḳ (A). Brit. Mus., MS. Add. 16,204, Folio 15*b*.

ገብርከ፡ወልደ
ሠማ ኒ ኒ ለሰዶር፡አ
ላዶር፡ዳናት፡አ
ዴራ፡ሮዳስ፡በሩ
ቅንዋተ መስቀ
ሉ፡ለእግዚእነ ኢ
የሱስ፡ክርስቶስ
ተማኅፀንኩ፡አነ
ገብርከ፡ኢስጢፋ
ኖስ

በስሟ፡አብ፡ወወ
ልድ፡ወመንፈስ፡ቅ
ዱስ፡፩አምላክ፡
ጸሎተ መንገደ
ሰማያ፡በሂት፡ም
ድር፡ዕቀበኒ፡ክር
ስቶስ፡ከመ አያ
ዕቅፍዋ፡ለነፍስ
የ፡መላእክተ ጽ
ልመት፡ከመ፡ት
ፊኑ፡ሊተ፡መላእ
ክተ፡ብርሃን፡ማ
ኅሊል፡ወገብር
ኢል፡እለ፡ግሩማ
ን፡ወጵሬቅሎጦ
ስ፡መንፈሰ ጽድ

Lefâfa Ṣedeḳ (A). Brit. Mus., MS. Add. 16,204, Folio 16a.

ቅ፡ ከመ፡ አ.ጸዕቅ
ፍዎ፡ ለዓለም፡ መ
ላእክት፡ ጽልመት፡
ወከመ፡ አ.ያቅሙ
ኒ፡ እግዚ.አ፡ ውስ
ተ፡ ጽልመት፡ ወ
የሐቂ፡ ስነነ፡ ተሞ
ዓፀንኩ፡ በ፡ግንጿ
ዊ፡ ስምከ፡ አነ፡ ገ
ብርከ፡ ወልደ፡ ማ
ካኤል ወበስማ፡ ለ
ማርያም፡ ድንግል
ወላዲት፡ አምላክ
ጥብርያዶስ፡ ወበ
አምልኮቱ፡ ለ
ስሟ፡ ያወ፡ ደን፡ ወለ

ማያ፡ ሰማያት፡ ወ
በመንበረ፡ ክብሐ
ተ፡ሁ፡ ዘሐነፀ፡ ጽር
ሐ፡ ወአልቦ፡ ዘየአ
ምሮ፡ ዘእንበለ፡ ባ
ሕቲቱ፡ ወዘእንበ
ለ፡ ክርስቶስ፡ ወል
ዱ፡ መሐሪ፡ መሐ
ርከ፡ ከ በለኒ፡ ወ
ስረዩ፡ ሊተ፡ ኃ
ጢአትየ፡ ለገብር
ከ፡ አስጠፋኖስ፡
ጠበ ደ፡ ኃሬት፡ ወህ
የ፡ ደንለሠ፡ ጊ፡ መ
ላእክት፡ ወ ጊ፡ መ
ንጦላዕት፡ ወያ

LEFÂFA ṢEDEḲ (A). Brit. Mus., MS. Add. 16,204, Folio 16b.

ዓር ጉ፡ጸሎተ፡በ
እንቱ፡ያመሕረቱ፡
ሰብእ፡ማኅጎቶቱ
መላእክቲሁ፨እ
ሙ፡ይቤሎ፡እግ
ዚአ፡መኑ፡ሰብእ
ዘአደአብኦ፡ወ
አደኑ፡ዕፅ፡ዘአ
ደጠደስ፡ውሉ
ዱ፡ሰብእ፡ዘአደ
ገብር፡ኃጢአተ
እግዚአ፡አልቦ፡
ኅር፡ዘእኀበለከ፡
ወሰበሃ፡ተናር
ሙ፡እግዚአብ
ሔር፡ለጊወዩሐ

ዋርደቱ፡ለርመ
ዌ አርዴ እቱ፡ከ
ሠመ፡ደጸሌፋ፡ዋ፡
ለዛቲ፡መጸሐፍ
ወደበሎሙ፡ለ
ሊዋርደቲሁ፨አ
ባህኩ ኩሙ፡ወዚ
ነው፡ለኵሉ፡ዘየ
አምን፡በየ፡በስ
መ፡አየሱስ፡ክር
ስቶስ፡ወልደ፡እ
ግዚአብሔር፡በ
ፁዕ፡ውእቱ፡ዘየ
አምን፡በየ፡በቃ
ለ፡ዝንቱ፡መጽ
ሐፍ፡ለዘጸሐፎ፡

ወለዘ፡አጽሐፎ፡ወ
ለዘ፡ዓነቆ፡በክሣ
ዱ፡ወበሰማየ፡ሂ
ሉ፡ተ፡ሂ፡ለእሙ፡
ተሐፅበ፡ወበበ
ቱሂ፡ለእሙ፡አኀ
በረ፡ጠዊ ቶ፡አያ
ጠውቶ፡ወየሐ
ዩ፡በደኃሪቶ፡ዕ
ለቶ፡አመበ፡ዕለተ
ኰነኒ፡ወደ ደኀ፡
ደቶ መሐረ፡መአ
ምሕሮ፡አምእሰ
ተ፡ገሃነም፡አመ
ፅለቶ፡ደቶሉ ለዩ
ኃጥኣን፡ወዓማፅ

ያኀ፡አመሂ፡መዓ
ልተ፡ወእመሂ፡ሊ
ሊተ፡ነበ፡ሀለወ
ቶ፡ወፃሮ፡በሁዕ
ወ፡አቴ፡አስሙ
ዓባያ፡ያ.አቲ፡ዕ
ለቶ፡እግዚአቲ፡
ዐበዴሊፋ፡አንተ፡
ቶ ሰዩ ዩ፡አጋንን
ቶ፡ወፃዕረ፡ጥቶ፡
አምሳዕለ፡ነኝሰ፡
ኀብረ ኮ፡ወሏ ዩ፡
ሀገ ክሌ ወተ ሰዩ ሻ
አምሳዕለ፡መጊበ
ረ፡ሰብሔተሁ፡ለ
እግዚአብሔር፡

LEFÂFA ṢEDEḲ (A). Brit. Mus., MS. Add. 16,204, Folio 17b.

ስዒሰመ፡ዓለም ተ፡ዘተሰትለ፡ወ
እሚ፡ኅቍበድ፡ማ ልደ፡ማር፡ያክባ፡ንዝ
ሂል፡ስጠ፡ኃደል ራ፡ዊ፡ጎጉ፡ሠ፡እየ፡
ከ፡ወበቸ፡ቢ፡ል፡ስ ሁ፡ድ፡ተዘከረነ፡
ምከ፡ወበልቅእ፡ እግዘ፡ኡ፡በወ፡ስተ
ል፡ስጠ፡ጥያጦተ መንግ፡ሠት፡ከ፡ለ
ተ፡ከ፡በጉ፡ሁ፡ካእ፡ ገ፡ብር፡ከ፡ወአ፡ሃ፡ማ፡
ል፡በዘፈ፡ታሔከ፡ ኡ፡ልሳጶር፡እስያ፡ር
እዓው፡ገተ፡ሰ፡ኡል ዴናት፡አየ፡ራ፡ሮ
በተተናዊ፡ወበሰ ዴስ፡በራ፡ትንጥተ
ተናዊ፡ወተር፡ነላ መስተለ፡ለእግዘ
ዊ፡ስምከ፡ተማ አነ፡ኢ፡የሱ፡ስ፡ክር
ዓዐነከ፡ከመ፡ት ስቶስ፡በዝነተ፡እ
ምሐረነ፡ወጉ፡ሣ ስማጊ፡ከ፡ተማ፡ዓ
ሃለኒ፡ለገብር፡ከ ዐነከ፡ወእጣዓ፡ዐ
እስወ፡ፋ፡ናስ፡ ዝየ ነከ፡ነፎ፡ስሂ፡ወሡ

Lefâfa Ṣedeḳ (A). Brit. Mus., MS. Add. 16,204, Folio 18a

ጋዩ፡ለገብርከ፡
ኤስጠፋኖስ ለዓ
ለመ፡ዓለም፡አ
ሜን፨ ፨ ፨

በስመ፡አብ፡ወወል
ድ፡ወመንፈስ፡ቅ
ዱስ፡፩አምላክ፡መ
ጽሐፈ፡አርድእት፡
ዘተሰአልዎ፡አር
ዳኢሁ፡ለኢየሱስ
እስከ፡ይትከሠት፡
ስሙ፡ኅቡእ፨ወኢ
ያምጽኡ፡ኣሬሁሱ፡ነገ
ርሙ፡ኢየሱስ፡ወ
ይቤሎሙ፡ዕተበ
ዎ፡ወአጽንዕዎ፡ወ
ትድኅኑ፡እምእሳ
ት፨ወክሎ፡ሰብ
እ፡ዘአእመረ፡አስ
ማተ፡ዘአጸንዖ፡

ወነገሮ፡ ወዘተሐ
ፀበ፡ አንበበ፡ እም
ብዝኃ፡ ኃጢአቱ
ደይሳንቶ፡ ዘጸሐፈ
አያመላከነ፡ በቃሉ፡
ወበእደዊሁ፡ ቅዱ
ሳት፡ ወመሀበኮ
ሙ፡ ለእርድእሁ
ከመ፡ ያንብብዎ
ወእንዘ፡ ያነብቡ
ረከቡ፡ ስሞ፡ ተ
ፈሥሑ፡ ወተሐ
ሠዩ፡ ወይቤሉ
እኑት፡ ወስቡሐ
ስምከ፡ ዘአርአይ
ከነ፡ ዘኖተ፡ ኰሉ፡

ወሀብከነ፡ ስመ
ከ፡ ቅዱሰ፡ ወጸ
ውዑ፡ ስሞ፡ አገ
ዘ፡ ደብሉ፡ ራፎ
ን፡ ራፎን፡ ራፎ
ን፡ ራኮን፡ ራኮን
ራኮን፡ ጹስ፡ ጹ
ስ፡ ጹስ፡ አፍሎ
ስ፡ አፍሎስ፡ አፍሎ
ስ፡ ምልዮስ፡ ም
ልዮስ፡ ምልዮስ፡
ሐናኤል፡ ሐናኤ
ል፡ ሐናኤል፡ ጽራ
ኤል፡ ጽራኤል፡
ጽራኤል፡ ናርስ
ናርስ፡ ናርስ፡ ከ

ሮስ፡ኪሮስ፡ኪ
ሮስ፡ፊሎስ፡ፊ
ሎስ፡ፊሎስ፡ሰ
ሮስ፡ሰሮስ፡ሰሮ
ስ፡ሰፋሮናስ፡ሰ
ፋሮናስ፡ሰፋሮና
ስ፡ኔሮን፡ኔሮን፡ኔ
ሮን፡ኤሮን፡ኤሮ
ን፡ኤሮን፨እምነ
ሎሙ፡ዘየዓበ
ስሞሙ፡ያሜሂል
ያሜሂል፡ያሜሂ
ል፡ብርስባሂል፡ብ
ርስባሂል፡ብርስባ
ሂል፡አትሜሂል
አትሜሂል፡አትሜ

ሂል፡አልቦ፡ዘየአ
ምሮ፡ለዝንቱ፡ስሙ
የ፡ዘ [illegible]
ክህናት፡ሰማያት፨
ወዘእንበለ፡ሠሮ
ፂሁ፡እምፂሁ፨ወፂአ
ሎሙ፡ [illegible]
ዝንቱ፡ስምየ፡ [illegible]
[illegible]
[illegible]
[illegible]
[illegible]
[illegible]
[illegible]
[illegible]
[illegible]

ለዩዩን፤ እምኵሉ፡
ዘተጽሕፈ፡ውስ
ተ፡መጻሕፍትየ፤
አልቦ፡ዘየዓብዮ
ለዘንቱ፡ነገር፤እ
ምኵሉ፡ጸሎት፡
በዝንቱ፡ዘተአ
መነ፡እምሕር፡
ወእሣሃሉ፨መ
ሀልኩ፡በመንበ
ርየ፡ወበርእስየ፡
ልቡል፨መሀል
ኩ፡በመከየደ፡
እገርየ፨ወበሣ
ርጸ ምእፃዐየ፨መ
ሀልኩ፡በቅዱሳ
ን፡መላእክትየ
አንሰ፡አይድዓም
ዕ፡በጽጹቅየ፡
ወሕይኑሱ፡በቱ
ልየ፨ወኤይረ
ኵስ፡ከዲንየ፡በ
ከመ፡አደኅንኮ
ሙ፡ለቅዱሳን፡
አርዳኢከ፨አያ
ኅነኒ፡በኃይለ፡ስ
ምከ፡ቅዱስ፡ሕ
ዕበኒ፡ወአንጽ
ሐኒ፡እምኃጢ
አትየ፡ለገብርከ፡
ዉልደ ማካኤ፨ወ
ካዕበ፡ይቤሉሙ

LEFÂFA ṢEDEḲ (A). Brit. Mus., MS. Add. 16,204, Folio 20a.

LEFÂFA ṢEDEḲ (A). Brit. Mus., MS. Add. 16,204, Folio 20b.

ሠራቂነ፡ አ.ይ.ክ
ል፡ሠረቶ፡ወጸላ
ኢ.ነ፡ አ.ይ.ክሉ፡ ወ
የደከምህ፡ኃይሳ፡
ኰሉ፡በሩ⁕ወይ
ትባረክ፡ቤቱ፡ወ
ውሉዱ⁕ወጠሳ
እክትነ፡ አ.ይርሳ
ቱ፡ እምነሆ፡ ወት
ረ፡በረከተ ነቢ
ያት፡ ወሐዋርያት
ወመንፈስ፡እግዚ
እብሔር፡የዐርፉ
ላዕሊሁ⁕ወመነ
ፈሰ ሰየጣነ፡ ይር
ኃቅ፡እምነሆ⁕ወ

እንተሃ፡ ለእመቱ
እመነክ፡ ዘነተ፡ ጸ
ሉተ፡ ወማየ፡ ጸሎ
ቱ ሃ፡ አ.ይትከባወ
ው ስተ፡ ምህርቁ
እስጠ፡ ክቡር፡ ወ
ቅዱስ፡ ውእቱ፡ አ
ምሳለ፡ ሥጋሁ፡ ወ
ደሙ፡ ለክርስቶስ
መንጾሔ፡ ኃጠአ
ት፡መህ ኃነ፡ ተ፡ ነ
ፕስ፡ወሥጋ⁕ወ
ዘነተ፡ አንበ.በከ፡
ለእመ፡ተሐጸብከ
ትመውዕ፡ወተገ
ርር፡ በረክ፡ ወጸሳእ

ተከ፨መአልቦ፡ዘይ
ክል፡ቀዊመ፡ቅድ
መ፡ገጹ፡ከ፡ኵሉ፡ፍ
ጥረት፡ይረዕዱ፡አ
ምቅልክ፨መሰበ፡ይ
ሬእዩ፡ገጸከ፡ይኄ
ይዩ፨ወይጥዕም፡
ነገርከ፡ለኵሉ፡ሰ
ብእ፨ተዘከረነ፡
እግዚኦ፡አመ፡ት
መጽእ፡በመንግ
ሥትከ፡ለገብርከ
እስጢፋኖስ፡
በስመ፡አብ፡ወወል
ድ፡ወመንፈስ፡ቅ
ዱስ፡፩አምላክ፡

ኢስምዓቱ፡ዘነገር
እግዚእነ፡ለአንጻ
ርዮስ፡ትዲስ፡ሬ
ደ፡አ፨ወይቤሎሙ፡
ሑር፡ሀገረ፡በላዕ
ተ፡ሰብእ፡ኀበ፡ሀ
ሎ፡እኁከ፡ማዓትየ
ስ፡ከመ፡ታውጽ
እ፡እምቤተ፡ሞ
ቅሀ፨ተንሥእ፡
ወሑር፡ምስለ፡
ዬ ሐርዲአከ፡ወአ
ውሥእ፡እንጻር
ያስ፡በእፎ፡እክል
በጸሎት፡ለይእቲ፡
ሀገር፨እስመ፡ር

Lefâfa Ṣedeḳ (A). Brit. Mus., MS. Add. 16,204, Folio 21b.

ኃትቱ፡ መጠነ፡ ዴ
ዓመፀቱ፡ ኢይክል
በጺሐ፡ ሶቤሃ፡ ዓ
ቢይ፡ ባሕር፨ ወ
ውስቴታ፡ ሀሎ
ወአውሥአ፡ እ
ግዚእ፨ ወይቤሎ
ኢትፍራህ፡ ኦእግ
ዚርያስ፡ ፍቁርየ
እከሥት፡ ለከ፡ ዓ
ቢየ፡ ነገረ፡ ወመ
ዲምጮ፡ ውስቴ
ቱ፡ ወእነግረከ፡ አ
ስማተ፡ ሶበ፡ ትበ
ጽሒ፡ ወትዴለው
ለሐዊር፡ ከመዝ።

በሌ፨ አግዞርያስ።
አርያስያስኖስ፡ አር
ያስያስኖስ፡ አርያ
ስያስኖስ፡ ኪያዩ
ዱዩስ፡ ኪያዩዳዩ
ስ፡ ኪያዩዳዩስ፡ አ
ክሌያዳኤል፡ አክ
ሌያዳኤል፡ አክሌ
ያዳኤል፡ ሰርኑኤ
ል፡ ሰርኑኤል፡ ሰር
ኑኤል፡ ተዳኦስ፡ ተ
ዳኦስ፡ ተዳኦስ፡ ር
ዴያኤል፡ ርዴያኤ
ል፡ ርዴያኤል፨ አ
ስማቲሁ፡ ለአበ
የ፡ እምቅዱ፡ መ፡

ገፍጥሮ፡ ለሰማ
ያ፡ ወምድር፡ አስ
ማቲሁ፡ እንግሪከ፤
ቀዲሙሰ፡ ዘነገረ
ኩከ፡ አስማቲሁ
ለሉብዩ፡ ውእቱ፡
ስምየ፨ ሰልጣዊ
ቲኤል፤ ሰልጣቄቲ
ኤል፤ ሰልጣዋቲኤ
ል፤ ፀበሮቱና ኤል
ፀበሮትና ኤል፤ ዐ
በሮትና ኤል፤ ቲዳ
ኤል፡ ቲዳኤል፡ ቲ
ዳኤል፤ አግስያዩስ
አግስያዩስ፡ አግስ
ያዩስ፤ ልምዩስ፡

ልምዩስ፡ ልያሁዩ
ስ፡ አስቲዲዮስ፡ አ
ስቲዲዮስ፡ አስቲ
ዲዮስ፨ ዘበትሮ
ሰሚሁ፡ ኢዮሱስኮ
ርስዮስ፡ ብሂል፤
ዲዲሚል፡ ዲዲ
ሚል፡ ዲዲሚል፤
አስሬል፡ አስሬል
አስሔል፨ አስማ
ቲ፡ መኀሬስ፡ ቅዲ
ስ፡ ጳራቅሊጦስ፤
አሬያዩል፡ አሬያ
ዩል፡ አሬያዩል፤
ዳኤል፡ ዳኤል፡ ዳ
ኤል፤ ኢሉሂ፡ ኤ

ሎሂ፡ ኤሎሂ፡ ጸ
ባዖት፡ አዶናይ፡ ጋ
ዮስ፡ ጋዮስ፡ ጋዮ
ስ፡ ኡጋዮስ፡ ኡጋ
ዮስ፡ አጋዮስ፡ ዘ
በትርጓሜሁ፡
ቅዱስ፡ ቅዱስ፡ ቅ
ዱስ፡ እግዚአብ
ሔር፡ ጸባዖት፡ ፍ
ጹም፡ ምሉዕ፡ ሰ
ማያት፡ ወምድረ
ቅድሳተ፡ ስብሐቲ
ከ፨ አልክናተ፡ ዘ
በትርጓሜሁ፡
ሃሌ፡ ሉያ፡ ለአብ፡
ሃሌ፡ ሉያ፡ ለወል

ድ፡ ሃሌ፡ ሉያ፡ ለ
መንፈስ፡ ቅዱስ፡
ስብሐት፡ ለአብ፡
ስብሐት፡ ለወልድ፡
ስብሐት፡ ለመንፈ
ስ፡ ቅዱስ፨ ለዘህላ
ዌሆሙ፡ ዕ̃ ኵሉ
ጊዜ፡ ኅቡረ፨ ይእ
ዜኒ፡ ወዘለፈኒ፡
ወለዓለመ፡ ዓለ
ም፡ አሜን፨ አል
ቦ፡ ዘ ተናገርኩ፡
ዘንተ፡ ነገረ፡ ወለ
ማርያምሂ፡ እምነ
ዮ፨ ወለክሰ፡ ከሠ
ትኩ፡ ወጸለዩ፡ በ

LEFÂFA ṢEDEḲ (A). Brit. Mus., MS. Add. 16,204, Folio 23a.

ዝንቱ፡ አስማት
የ፨ወትርጓሜሁ፡
አንቅጽ፡ ወትራት
ሔ፡ ሙቁሔን፨ወ
ዘንቱ፡ አስማት፡
ለእሙ፡ ሃሮ፡ ወሚ
ኑቶ፡ ደኪመ፡ ገ፡ ክፋ
ሉ፡ ዓመሰለ፡ ጴጥሮ
ስ፡ ሊቀ፡ ሐዋርያ
ት፤ አይሬእዩ፡ ዓይ
ን፡ እኩይ፡ ወእይቀ
ርቦ፡ ኃይለ፡ ጸላዒ፤
ወአረዋጺ፡ ኃይለ
አጋንንት፡ እኩያን
ወኃይለ፡ መናፍ
ስት፡ ርኩሳን፤ ሊ
ደኪሉ፡ ኃይለ፡ ጸ
ለመት፨ ጊሬዮን
ሚሊሱ፡ ገደን፡
ሰተናጥ፡ ቀተናጥ፡
ተንክራምን፡ ቀታ
ሊ፡ ወሚሃየዋ፡ ክ
ርስቶስ፡ መልደ፡ እ
ግዚአብሔር፡ ወ
መልደ፡ እግዚአት
ነ፡ ማርያም፡ ዘአሰ
ረነ፡ ለብርያል፡ ክ
ማሁ፡ አስሮሙ፡
ለዐርየ፡ ወለጸላኢ
ትየ፨ ተዘከረኒ
እግዚኦ፡ በኃይለ፡
ዝንቱ፡ አስማት፡ ክ

LEFÂFA ṢEDEḲ (A). Brit. Mus., MS. Add. 16,204, Folio 23b.

አመ፡ትመጽእ፡
በመንግሥትከ፡
አነ፡ገብርከ፡ወል
ደ፡ሚካኤል፡በ
ስመ፡አብ፡ወወል
ድ፡ወመንፈስ፡ቅ
ዱስ፡፩አምላክ፡
አስፀዓተ፡እግዚእ
ነ፡ኢየሱስ፡ክርስ
ቶስ፡ሰድራሳዊ፡ክ
መ፡ትምጽአኒ፡
ሞት፡ዘእንበለ፡ጊ
ዜየ፡በአውላኬት፡
በድርዳስ፡በናርኖ
ስ፡በኢሉኝ፡በደ
ልፎጊን፡በጋዲን፡

በዮጣ፡ወቢበኩ
ላዲን፡በሰዶራቃ
ኢል፡ዘቂሬቂጢ
ን፡በዶሎቆሎን፡
በዘሩባኤል፡በጹ
ፏፋኢል፡በዶሎ
ሆሎሂን፡በቆለኪ
ን፡በከፋዚን፡በጋ
ዚን፡በፏላክኢል፡
በአልፋኤል፡በዳ
ራኋን፡በዝራኢ
ል፡በገልማሣሳዊ፡በ
ገለውዱያን፡በኢ
ደፌን፡በቀላዲን፡
በአብዳዊ፡በም
ናሲሳዊ፡በስልና

LEFÂFA ṢEDEḲ (A).   Brit. Mus., MS. Add. 16,204, Folio 24a.

ኵስ፡በየ፡ሳዊ፡በነ	ል፡እ፡ያእ፡ል፡ሃ፡ያ
ለዳፈ፡ን፡በቀላኢ፡	እ፡ል፡ሃ፡ዳእ፡ል፡ሃ
ል፡በየፉኢ፡ል፡በ	ዳእ፡ል፡ሃ፡ዳኢ፡ል
ስድራታኢ፡ል፡በሲ	ሃ፡ዳኢ፡ል፡ሃ፡ዳኢ
ለ፡ተማዓዐነከ	ል፡የ፡ሀ፡ናኢ፡ል፡ሀ
እነ፡ንብርከ፡እሰመ	ህ፡ናእ፡ል፡የ፡ሁ፡ና
ፈኖስ፡ በእመ፡እ	ኢ፡ል፡የ፡ሁ፡ና፡እ፡ል
ብ፡ወወልህ፡ወ	የ፡ሁ፡ናእ፡ል፡የ፡ሁ
መነፈ፡ስ፡ቅዱ፡ስ	ናእ፡ል፡የ፡ሁ፡ናኢ
ቅ፡እግዚአከ፡እል	ል፡እ፡ርናኤ፡ል፡እ
ፋ፡ወእ፡እልፉ፡	ርናኢ፡ል፡እ፡ርናእ
እልፉ፡እልፉ፡እ	ል፡እ፡ርናኢ፡ል፡እ
ልፉ፡እልፉ፡እል	ርናእ፡ል፡እ፡ርናእ
ፉ፡እልፉ፡ኢ፡የ	ል፡ሁ ርናእ፡ል፡ሃ
እ፡ል፡እ፡ያእ፡ል፡እ	ርናእ፡ል፡ሃ፡ርናእ
ያእ፡ል፡እ፡ያእ፡ል	ል፡ሃ፡ርናእ፡ል፡ሃ
እ፡ያእ፡ል፡እ፡ያእ	ርናእ፡ል፡ሃ፡ርናእ

ል፡ሂርና፡ኢ.ል፡ሂር ል፡እክሃል፡ሩ-ት፡ኢ.
ና፡ኢ.ል፡እማ.ስ፡እ ል፡ተ፡ቲ፡ት፡ሃል፡ርት
ማ.ስ፡እማ.ስ፡እማ. ሃል፡እ.ል፡ሃል፡ቲ፡ታ
ስ፡እማ.ስ፡እፀሃ.ስ እል፡ሆ፡ል፡ሃል፡ክር
እማ.ስ፡ ፡ሃ ቲ፡ሃል፡ሰ፡ብ፡ት፡ሃል፡
ህ፡ሃ፡ገዓ፡ሄ፡ክ፡ነ፡ ማ፡ታ፡እል፡ማ.ራ.
ህ፡ሀ፡ሃ፡ዒ፡ ስራ፡ሃስ፡ሃ ኡ.ል፡እክ፡ስ፡ሩ፡እ.ል፡
ል ስ-ር፡ሃ.ል፡ሩ፡ር፡ሄ-

LEFÂFA ṢEDEḲ (A). Brit. Mus., MS. Add. 16,204, Folio 25a.

ቲር፡ናስከ፡ብ፡እከ ፡ የሀል፡አር፡ምየሀል፡
ስኑ፡የከ፡አ፡ናር፡ እርየዋሂ፡አናምየ
በራ፡ከ፡የከ፡ራ፡ስቦ፡ ል፡አልሂ፡ የል፡አዉ
ምን፡ዊከ፡የስ፡እርነ የዕደ፡ እ፡ዘ፡ፈ፡ሳማ፡
የስ፡ጥራ፡ስ፡ክ፡ናስ፡ ፈ፡የማ፡ሰርሂ፡ር፡መ
አበጸሉ፡ን፡እንስከ፡ ትጥሂ፡ሂ፡ እራ፡ሂ፡የ
ወ፡ሂ፡ምጠከ፡ወ ል፡ረ፡ወ፡ር፡ፈ፡ሩል፡
ት፡እክትና፡ለ፡እጸ ፡ ፍርት፡ክ፡ሉ፡ሀለ፡ሉ
ከ፡ና፡እራ፡ጸ፡እንየ ሀለ፡ሉ፡ሀለ፡ መ ፡ ክ
ከ፡ሰር፡ሂ፡ ክላሰ፡ን ኢ፡ል፡ወ፡ን፡ብር፡እ፡ል
እ፡ሰ፡ራ፡ን፡ ማ፡ ራ̂፡ከ፡ ሉ፡ር፡የል፡ወስሂከ፡የ
ጥር፡ከ፡ወር፡ዲ፡እከ፡ ል፡ሰራ፡ት፡የል፡ወ፡እ
እጸ፡ማ፡አል፡ከ፡ና፡አል፡ ን፡ን፡የል፡ ራ፡ፊ፡ ኢ፡ል፡
ሂ፡ራ፡ን፡ እር፡ና፡ ማ አ ራ፡ጥ፡ የል፡ወ ኃ
ረከ፡ስስን፡ከ፡እማ፡ ር፡ማስ፡የል፡እት፡ማ
የ፡ስ፡የ፡ወ፡ ረ፡በር፡ ሂ፡የል፡ እ፡ዲ፡ ምየ
ሂ፡ምየል፡ማስ፡ሂ፡ ል፡ እር፡ን፡የን፡ል፡ እስ

Lefâfa Ṣedeḳ (A). Brit. Mus., MS. Add. 16,204, Folio 25b.

ሩፃሙ፡ዚዲኦሌ፡ሱ
ሩክ፡ፃመንሱክ፡አፋ
ብርያኖስ፡ኪሩቢ
ል፡አፍኒንያል፡አት
ልዋ፡ብርስትያል፡
ኡብርያል፡ኡብሩቅ፡
ሩጣ፡ፏርትያል፡ፋ
ርፉር፡ፉማቀቀል
ፋኖንያል፡ዲዩያ
ል፡መሬዩክዮል፡
ኡፋዩክያል፨አሥ
ሉስ፡ቅዱስ፡ተማኅ
ሀንኪ፡በበአስማ
ቲክሙ፡ወበስሙ፡
መላእክቲክሙ፡ወ
ካህናቲክሙ፡ንሙ
አ ቅሬቡ፡በየማ
ንየ፡ወበበኃምየ፡
በቅዱሚዩ፡ወበዱ
ኅሬዩ፡ጠኖኑስተ፡
ርኩሳን፡ጠሠራዊ
ተ፡ዲያብሎስ፡እ
ኩያን፡ኀበሀሎኩ
አነ፡ገብርክሙ፡ሠ
ሉዩ ማገካኢ፡ኡዩስዩ
ን፡ርዩኑ፡ሣዱሬ፡
ኡኡሱሱኖያክክ
አዮዎስ፡ሰላስኢል
ሂሲዎኚ፡ዴንፍስ፨
እሰጠ፡ፉ ኖሰ

ተማኅፀንኩ፡ አነ፡
ገብርከ፡ ኤልያስ
ኪዱ፡ ፡ወፈድፋ
ደሰ፡ ነፍስ፡ ገብርከ
እገሌ። በ፸፻ቢጽ
በ፫፻፲ወ፲ርቱዓ
ነሣዲ ማኖት፡ ዘተ
ማኅፀንት። ፍት
ሐ፡ እግዚኦ፡ አማ
ፅሠረ፡ ኃጢአት፡
ለዓለመ፡ ዓለም።
አሜን። ። ።
በስመ፡ አብ፡ ወወ
ልድ፡ ወመንፈስ፡ ቅ
ዱስ፡ ፩አምላክ።
ዛቲ፡ ጸሎት፡ መድ
ሐፊ፡ ሕይወት፡ ል
ፋፊ፡ ጽድቅ፡ ዘጸ
ሐፈ፡ አብ፡ እምቅ
ድመ፡ ይት ወሰድ፡
ክርስቶስ፡ እማር

ያም፡ እንተ፡ ታበው
ዕ፡ ውስተ፡ ሕይወ
ት፡ ወውስተ፡ ጸባበ
እንቀጽ፡ ወታበጽ
ሖ፡ ውስተ፡ መንግ
ሥተ፡ ሰማያት፡ መ
ርህ፡ ለጻድቃን፡ ወ
ዘገተ፡ ነገራ፡ ሰማ
ርያም እምድኅረ፡
ተወልደ፡ እምኔሃ፡
አመ፡ ፳ወ፪ ለመጋ
ቢት፡ አርአየ፡ ክር
ስቶስ፡ ለማርያም
ኃጥአን፡ በውስተ፡
ደይን፡ መንበ፡ ደነ
ብሩ፡ ጻድቃን፡ በው
ስተ፡ ገነት። ወት
ቤ፡ ማርያም፡ ሰበ፡ ት
ሬኢ፡ ደንገፀት፡ መ
ርዕደት፡ ወፈርሃ
ት፡ ዐቢየ፡ ፍርሃተ።

ወይቤላ፡ ኢየሱስ፡
ኢትፍርሂ፡ ማርያ
ም፡ እምየ፡ እንተ፡
ፆርክኒ፡ በከርሥ
ኪ፡ ወወለድክኒ፡
በመንፈስ፡ ቅዱስ፡
ወካዕበ፡ ተሰአሎ
ወትቤሎ፡ ንግረ
ኒ፡ ወልድየ፡ በም
ንትኑ፡ አዝማድየ፡
እምዝንቱ፡ በሳዓ
እሳት፡ አንሰ፡ እፌ
ርሁ፡ በእንተ፡ ነፍ
ስየ፡ በእንተ፡ እም
የ፡ ወበእንተ፡ ኢያ
ቄም፡ አቡየ፡ ወበ
እንተ፡ ሳሙኤል፡
እኁየ፡ ወበእንተ፡
ኤልሳቤጥ፡ እኅት
የ፡ ወበእንተ፡ ጸዋ
ት፡ ዘመድየ። ወይ

ንዚኒ፡ ኗጣሪኒ፡ ዋ
ቀ፡ በዘይጵሳኑ
አፃዝኗቱ፡ አሳት፡
ወይቤሳ፡ ኢየሱ
ስ፡ ኢያንጣረኪ፡ አ
ሰመ፡ ዘተናገረ፡
ኪ፡ ይወጽአ፡ ኀበ፡ ሣ
ለሳይ፨ ወአምጽ
ኣረሰ፡ ይዘራአ
ሁስተ፡ ክርስ፡ ስብ
አ፡ ወይገብሩ፡ ኂ
ጠኢተ፨ እኗዘ፡ ይ
ብሉ፡ ወሀሰወኒ፡
በዘኗጽኀኗ፨ ወ
ነዕበ፡ ተሰአሰቶ
ወተቤሎ፡ በእኗ
ተ፡ ምኗቱ፡ ፃርኩ
ከ፡ በክርሥየ፡ ፱
አውራኂ፡ ወ፳ዕ
ስተ፡ ወበከየት፡
ማርያም፡ እኗብዓ

...ፈረ፨ ወክርስ
ቶስኒ፡ በከየ፡ ም
ስሴሃ፡ ወይቤሳ፡
ኢየሱስ፡ ኢትብ
ክዩ፡ እምዮ፤ ናሁ
እነግሮ፡ ለአቡየ፡
ወለእመኒ፡ ወሀ
በኒ፡ እነግረኪ፡
ወይቤሎ፡ ለአቡ
ሁ፡ ማርያም፡ እምየ
ትበኪ፡ ሀበኒ፡ መ
ጽሐፈ፡ ሕይወት፡
እኗተ፡ ጸሐፍከ፡ በ
አጼከ፡ ቅድስት፡ አ
ምቅድመ፡ አትወ
ሰድ፡ አነ፡ ተነብሮ፡
በሠረገላ፡ ኪሩቤ
ል፡ መኗበርከ፡ ወ
ይቤሎ፡ አቡሁ፡ ለ
ወልዱ፡ ናሁ፡ ወሀ
ብኩከ፡ ኗጣሬ፡ ለ

ማርያም፡ እምከ፡ አ
ል [illegible] ዕኩ፡
እምኔከ፡ አሳ፡ ከሠ
ተኩ፡ ሰከ፡ ወጼ
ሐፈ፡ ኢየሱስ፡ በቀ
ለመ፡ ወርቅ፡ ወመ
ጽአ፡ ደመና፡ ብሩ
ህ፡ ወከሰሎሙ፡ ሃ
ጠኗጦሳስተ፡ ዘነደ፡
አሳት፡ ወአልቦ፡ ዘ
አአመሮ፡ ዘሰምዓ፡
መሳኒክተ፡ ወሰፃ
ነ፡ መሳእብት፡ አለ
ቦ፡ ይነግሮ፡ ዘኒተ
ነገረ፡ ወይቤሳ፡ ኗ
ሥአ፡ ዘወሀበከ
አቡየ፡ ሰማያዊ፡
ወለኗቲኒ፡ አ
በሥቴ፡ ለዘአዩ
በሌ፡ ዐዋሮቶ፡ ፴
ዐቀበሆ፡ ለዝኗቱ

LEFÂFA ṢEDEḲ (B). Brit. Mus., MS. Orient. No. 551, Folio 27*b*.

ነገር፡እንጻእ፡ሰጠ
ቢባኒ፡እሰ፡ሃአምኑ
ኀብየ፡ወእሰ፡የሐው
ሩ፡በትእዛዝየ፡ወ
ዘአጥረ ኜ፡ሰዛቲ፡
መጽሐፍ፡ኢይወ
ርድ፡ውስተ፡ደይኒ
ወሙስተ፡ሲኦል፡
ወሰእመኒ፡ፃሮ፡ወ
አነቆ፡በክሣዱ፡ይ
ትኃደግ፡ኃጢአቱ፨
ወሰእመኒ፡ደገመ
በቃሉ፡ሰሰቍርባ
ኒ፡ይነጽሕ፡እምር
ስሐተ፡ኃጢአት፨
ወሰእመኒ፡ገብሩ፡
ኀበ፡መጣነዙ፡ማዕ
ተበ፡ሰሱሞኒ፡ርበ
ዝኒቱ፡መጽሐፍ፡
ሰአመ፡ተቀብረ፡ይ
መርሀዎ፡መሳእከ

ት፡ውስተ፡እንቀ
ጸ፡ሕይወት፡ወያ
በጽሕዎ፡ቅድመ
እግዚአብሔር፡ወ
ያበውስዎ፡ውስተ
መንግሥተ፡ሰማ
ያት፡ወውስተ፡ብር
ሃን፡ሕይወት፡ወ
መድኃኒት፡ወመ
ድኃኒት፡ዘሰዓሰ
ም፨ወዘንተ፡ፈጺ
ሞ፡እግዚእ፡ኢየሱ
ስ፡ነገረ፡እስማቲ
ሁ፡ሰድልው፡ስሕ
ይወት፡ወሰመድ
ኃኒት፨ወከዕበ፡ይ
ቤሳ፡እግዚእ፡ኢየ
ሱስ፡ሰማርያም፡ወ
ይቤሎ፡በብርናኤ
ል፡ስምከ፡ተማኅ
ፀንኩ፡ከመ፡ትም

ሀረኒ፡ወትሠሃሰ
ኒ፡ወትደምሰሰ
መጽሐፈ፡ዕዳየ፡ሰ
ተ፡ሲተ፡ሰገብርከ
ተክለ ማርያም
በኢየሱስ፡ክርስቶ
ስ፡ስምከ፡ተማኅፀ
ንኩ፡ከመ፡ትምሐ
ረኒ፡ወትሠሃሰኒ
ወትምሰስ፡መጽ
ሐፈ፡ዕዳየ፡ሲተ፡
ሰገብርከ፨እሉቱ
ሉክቴ፡ በኪሮስ
ስምከ፡ተማኅፀ
ንኩ፡በጥርኮስ፡
ስምከ፡ወይቤሳ፡
ኢየሱስ፡ሰማር
ያም፡ፍጹም፡ምሕ
ረቱ፡ሰአቡየ፡ሰማ
ያዊ፡ወሰእመኒ፡ተ
አመኑ፡በዝንቱ፡

LEFÂFA ṢEDEḲ (B). Brit. Mus., MS. Orient. No. 551, Folio 28a

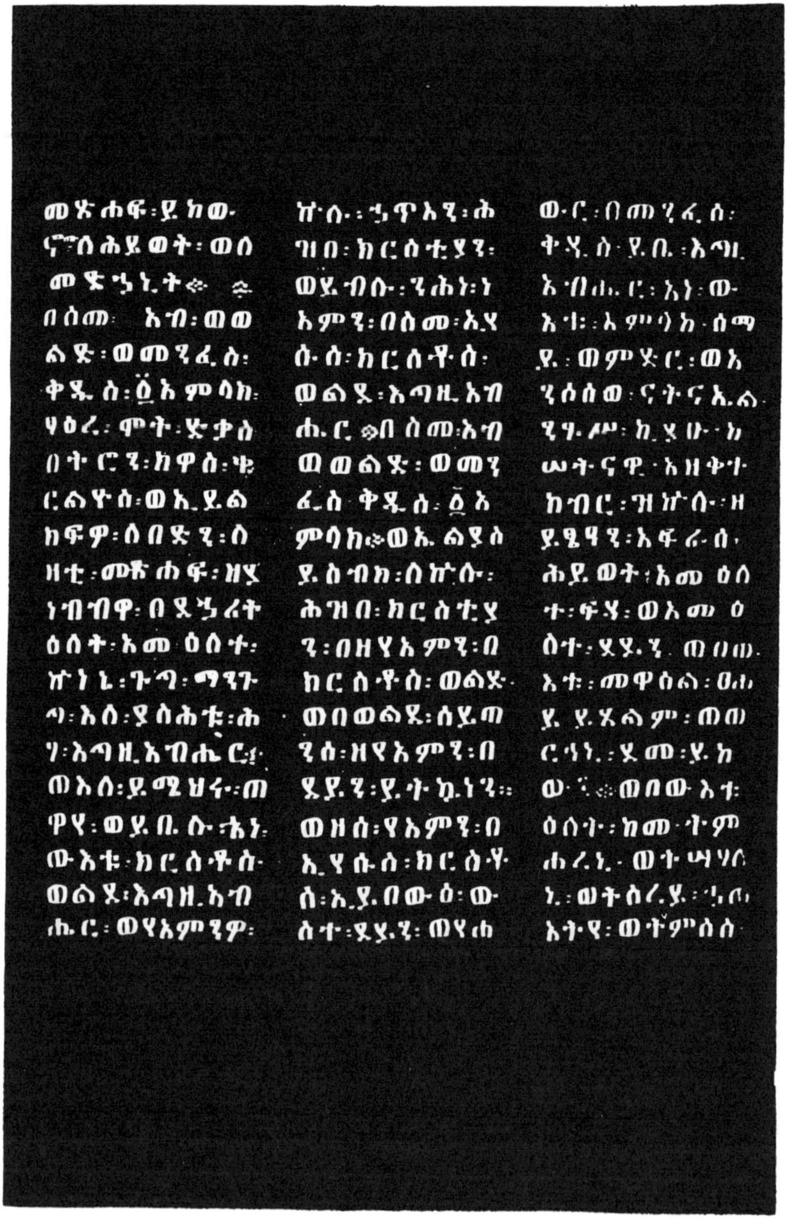

መጽሐፍ፡ ዘክው
ነሰሕይወት፡ ወሰ
መድኀኒት።
በስመ፡ አብ፡ ወወ
ልድ፡ ወመንፈስ፡
ቅዱስ፡ ፩አምላክ፡
ፃዕረ፡ ሞት፡ ድቃስ
በትሮኒ፡ ክዋስ፡ ቍ
ርልዮስ፡ ወኢይል
ክፍዎ፡ ሰበድኒ፡ ስ
ዘቲ፡ መጽሐፍ፡ ዘኗ
ነብብዋ፡ በጸኀሬት
ዕለት፡ አመ ዕለተ፡
ሖነኒ፡ ጉግ፡ ማኅጉ
ላ፡ እሰ፡ ያስሕቱ፡ ሕ
ገ፡ እግዚአብሔር፡
ወእለ፡ ይሜህሩ፡ ጠ
ዋየ፡ ወይቤሉ፡ ኤነ፡
ውእቱ፡ ክርስቶስ፡
ወልደ፡ እግዚአብ
ሔር፡ ወየአምኒዎ፡

ዘሰ፡ ኅጥአኒ፡ ሕ
ዝበ፡ ክርስቲያኒ፡
ወይብሉ፡ ኒሕነ፡ ነ
አምኒ፡ በስመ፡ ኢየ
ሱስ፡ ክርስቶስ፡
ወልደ፡ እግዚአብ
ሔር፡ ወበስመ፡ አብ
ወወልድ፡ ወመን
ፈስ ቅዱስ፡ ፩አ
ምላክ፡ ወኢልያስ
ይሰብክ፡ ሰኵሉ፡
ሕዝበ፡ ክርስቲያ
ኒ፡ በዘየአምኒ፡ በ
ክርስቶስ፡ ወልድ፡
ወበወልድ፡ ሰይጣ
ኒሰ፡ ዘየአምኒ፡ በ
ጸያኒ፡ ይትኳነኒ።
ወዘሰ፡ የአምኒ፡ በ
ኢየሱስ፡ ክርስቶ
ሰ፡ ኢይበውዕ፡ ው
ስተ፡ ደያኒ፡ ወየሐ

ውር፡ በመንፈስ፡
ቅዱስ፡ ዘበ፡ እግዚ
አብሔር፡ አነ፡ ው
እቱ፡ አምላክ፡ ሰማ
ይ፡ ወምድር፡ ወአ
ነሰሰወ፡ ናትናኤል
ኒሣሥ፡ ክዴሁ፡ ክ
ሠትናዊ፡ አዘቅት
ክብር፡ ዝሖሰ፡ ዘ
ይዬዋኒ፡ አፋሪስ
ሕይወት፡ አመ ዕለ
ተ፡ ፍኖ፡ ወአመ ዕ
ለተ፡ ደያኒ፡ ወበው
እቱ፡ መዋዕል፡ ፀሐ
ይ፡ ይጸልም፡ ወወ
ርኅኒ፡ ደመ፡ ይክ
ው። ወበውእቱ፡
ዕለት፡ ከመ ትም
ሕረኒ፡ ወትሣህረ
ኒ፡ ወትስረይ፡ ኃጢ
አትየ፡ ወትምስስ

መጽሐፈ፡ እጻዌ፡
ሲተ፡ ሰገብሮከ፡ እ
ግሌ፨ ስብሐት፡ ለ
እግዚአብሔር፡
በሰማያት፡ ወሰላ
ም፡ በምድር፡ ለ
ዘፈለጠ፡ ብርሃነ፡
ሂረድአነ፡ እምሳ
ዕነ፡ ወመድኃኒነ፡
ሰመከ፡ እኜዘ፡ ኜዜ
ከር፡ ወበመስቀ
ልከ፡ እኜዘ፡ ኜትእ
መኜ፤ ወበዓቡ ዕ፡
ስምከ፡ ኜሴብሐከ
ሰገዐሳኜ፡ ወሰል
ሃቀኜ፡ ከመ፡ ትም
ሐሬኒ፡ ወትሠሃለ
ሲኒ፡ ወትስረ ፎ
ኋጢአትየ፡ ወት
ምሰስ፡ መጽሐፈ
እጻዌ፡ ሲተ፡ ሰገብ

ርከ፡ ተኣለ፡ ማርያ
ም፡ ወያቤሎሙ፡
መልአክ፡ እግዚአ
ብሔር፡ ምኑትኑ
መ፡ ኑመ፡ ኵምፀ፡
ነጕድጓድ፡ ዘሰማ
ዕኩ፡ ወያቤሎ፡ እ
ግዚአብሔር፡ ለሚ
ካኤል፡ በጸጋየ፡ ው
እቱ፡ መኜበሮሙ፡
ለኃጥአን፨ እሰ፡ አ
ደግብሩ፡ ፈቃዷ፡ አ
ቡየ፡ ኣልቀተ፡ ነፍ
ሶሙ፡ ለእለ፡ ኢተዘ
ከርዋ፡ ቃሎሙ፡ ዘ
ይቤሎሙ፡ አመ፡
ዕለተ፡ ፋጻ፡ ወአመ
ዕለተ፡ ጸያኜ፡ ብፁዕ፡
ውእቱ፡ ዘአጽሐ
ፎ፡ ብፁዕ፡ ውእቱ፡
ዘአነቶ፡ በተአምኖ፡

ወኢያሰከፍዎ፡ መ
ሳእከት፡ ጽልመት፡
ሰበድኑ፡ መዋዕለ፡
ፀሐያ፡ ኢየየሮብ፡
ወማሓቶት፡ ዘኢያ
ጠፍዕ፡ ቃለ፡ ፍጻሞ
ሙ፡ ለኃጥአን፡ ዘኢ
ያረምም፡ ስብሐተ፡
መኜባሥቱ፡ ዘአያ
ትነስት፡ መኜበሩ፡፡
በአሳት፡ ከሱሉ፡ ዘ
አያኜቶስቅል፡ ለየ
ለመ፡ የሰም፨ ወያ
ቤልዎ፡ መሳእከተ፡
ሁ፡ ኜዜኑ፤ ስመከ፡
ከመ፡ ኜሴብሐከ፡ ወ
ኜዜምር፡ ስስምኩ፡
ወያቤሎሙ፡ ለቀ
ጻማዊ፡ ያአጹ፡ ካል
ዕ፡ ስምየ፡ ኪኜያ፡ ሣ
ልስ፡ ስምየ፡ አማኑ

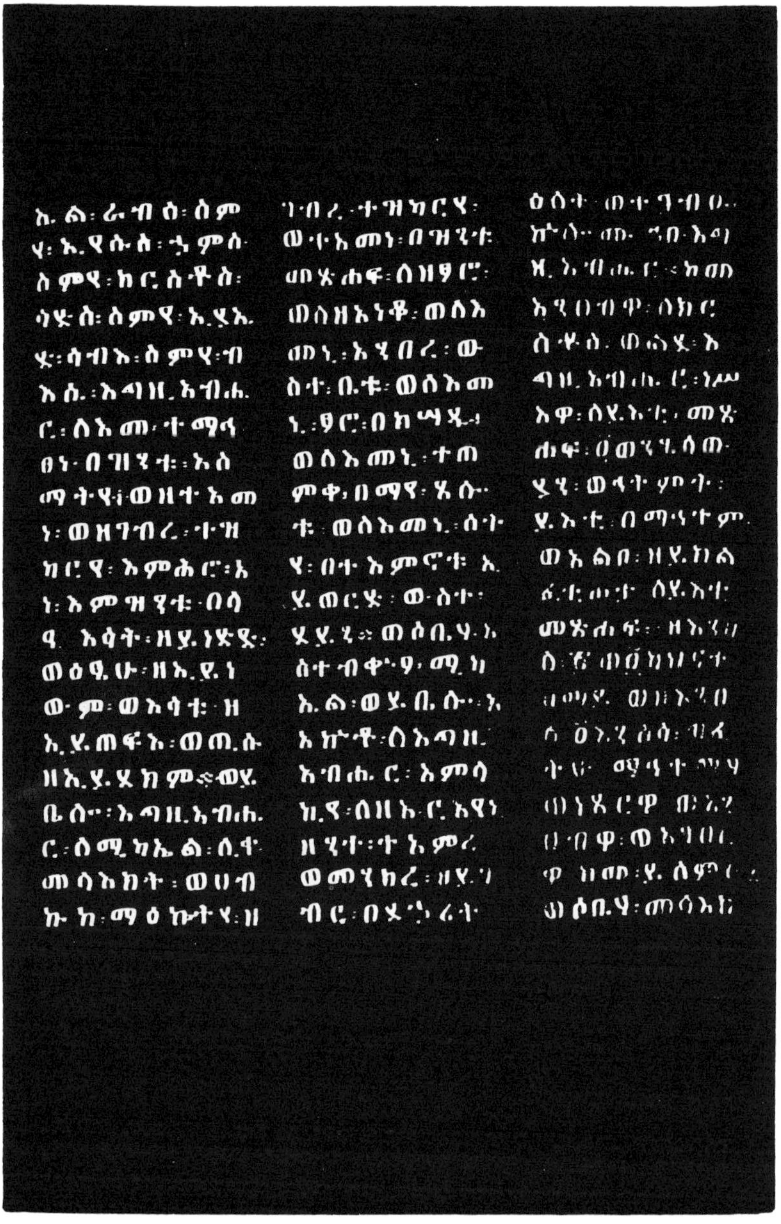

ኢ.ል፡ ራብ ሰ፡ ስም
ሃ፡ ኢየሱስ፡ ኀምስ
ስምሃ፡ ክርስቶስ፡
ሳድስ፡ ስምሃ፡ ኢያኢ
ዱ፡ ሳብእ፡ ስምሃ፡ ብ
እሴ፡ እግዚ አብሔ
ር፡ ለእመ፡ ተማኅ
ፀነ፡ በዝንቱ፡ አስ
ማቲሃ፡ ወዘተ እመ
ነ፡ ወዘገብረ፡ ተዝ
ካሪሃ፡ እምሕሮ፡ እ
ነ፡ እምዝንቱ፡ በሳ
ባ እሳት፡ ዘያነድድ፡
ወዕፁሁ፡ ዘኢይነ
ውም፡ ወእሳቱ፡ ዘ
ኢይጠፍእ፡ ወጢሱ
ዘኢያኀልቅ ምድ፨ወያ
ቤሎ፡ እግዚአብሔ
ር፡ ለሚካኤል፡ ሰ.ላ.
መሳእክት፡ ወሀብ
ኩከ፡ ማዕኩትየ፡ ዘ

ገብረ፡ ተዝካርየ፡
ወተአመነ፡ በዝንቱ፡
ሠበጹ ሐፍ፡ ለዘፃሮ፡
ወለዘአንቆ፡ ወለእ
መነ፡ አኒበረ፡ ው
ስተ፡ ቤቱ፡ ወለእመ
ነ፡ ፃሮ፡ በክሣዱ፡
ወለእመነ፡ ተጠ
ምቀ፡ በማየ፡ ሄሱ
ቱ፡ ወለእመነ፡ ሰተ
የ፡ በተ እምሮቱ፡ ኢ
ይወርድ፡ ውስተ፡
ደይኒ፨ ወሰቢሃኒ
ስተብቍዓ፡ ማኅ
ኢል፡ ወጌቤሉ፡ ኢ
አኰቴ፡ ለእግዚ
አብሔር፡ እምሳ
ኪየ፡ ለዘ አርአየነ
ዘኒተ፡ ተአምረ
ወመኒክረ፡ ዘየኀ
ብሮ፡ በጾኀራተ

ዕለተ፡ ወተገብሀ
ኵሉ፡ ሠራዊተ እግ
ዚ አብሔር፡ ከመ
[illegible]
[illegible]
[illegible]
አዋ፡ ለይእቲ፡ መጽ
ሐፍ፡ [illegible]
[illegible]
ይእቲ፡ በማኅተም
ወአልቦ፡ ዘይክል
[illegible]
መጽሐፍ፡ ዘእንበ
[illegible]
[illegible]
[illegible]
[illegible]
[illegible]
[illegible]
[illegible]
[illegible] ሶበሃ፡ መሳእክ

LEFÂFA ṢEDEḲ (B). Brit. Mus., MS. Orient. No. 551, Folio 29b.

ት፡ወነሥኡ፡፯መ
ላእክት፡መጥቅዓ፡
ወነሥኡ፡፯መላእ
ት፡ጽዋዓ፡ወከዓወ
ው፡ስቴ፡ገጾ፡መካኑ
ሃሞድር፡ከመ፡ይት
ቀደሱ፡ውሉደ፡ኄ
ር፡ወይትፌ ለጡ፡
ሰማየ፡፯ማዕዑት፡
ወ፯ብርሃናት፡ወ
፯መሣውሬ፡መናበ
ሬ፡መንበሩ፡ለእግዚ
አብሔር፡በዘቲአ
ምሩ፡ስሞ፡ግሩመ፡
ወኀበ ሀሰወ፡ነበያ
ት፡ወሐዋርያት፡በ
ሂብረ፡ልዑል፡ወበ
ኢብሬ፡መካኛ፡አመ
፲ወ፯ ሰመ ስክረም
ቀኌ ስ፡ሥጋሃ፡በኛጽ
ሕ፡በእኛት፡መስቀሉ

ለክርስቶስ፡ክቡር
ወመቃብሪ፡ክርስ
ቶስ፡ደስተርኢ፡ም
ሕረቱ፡በከመ፡ቃል
ከ፡ቅዱስ፡ወትበ
ሎሙ፡ስቅጻሳነ
ከ፡በፈሰሩ፡ከበ
ራት፡ ቀሉ፡ለእ
ግዚአብሔር፡አ
ግፎራ፡ግካኤል
ዝምራኤል፡ድና
ኤል፡ኒናት፡ዘ
ብጹ ምስ፡አማኑ
ዎስ፡መሰታግራቦ
ቶ፡ከመ፡ሄትር፡
ነፋያኖስ፡አሮራ፡
ቀሰ፡ቱዋራ፡ወር
ያኤል፡ኤልያና፡
ስሙ፡አቱዊ፡አቱ
ዊ ሳ፡ሳሰር፡ወከመ
ግዘ፡በፍክሬሁ፡በ

ግዕዝ፡ ወዝኒቱ
ዕርገቱ፡ለማርያ
ም፡ወከመዝ፡እዕ
ርገኒ፡ለገብርክ፡
ተ ኒለ ማርያም፡
፩ስሙ፡ሳፋዮስ፡፪
ስሙ፡ቆሆሆኬ፡፫
ስሙ፡ገብርኤል፡
፬ብርሃናኤል፡፭ስ
ምናኤል፡፮ስሙ
ገዘራኤል፡፯ስሙ
ስምርጻኤል፡፰ስ
ም፡ድድያ፡ወዝኒ
ቱ፡ኢሀሰወት፡በ
ልበ፡ሰብእ፡መዋቅ
ያኒ፡ዘተ፡ዘወፅአ
ት፡እምአፉሆሙ
ወእምቃሰ፡ክና
ፋሪሆሙ፡ለአብ፡
ወወልድ፡ወመንፈ
ስ፡ቅዱስ፡ለዝኒቱ፡

ስሙ፡ አጣዩ ሱ፡ ፩ስ
ሙ፡ አርጎሮና፡ ፩ስሙ
በተርጓ፡ ፩ስሙ፡ ጥ
ና፡ ፩ስሙ፡ ማስያ
ር፡ ፩ስሙ፡ መዱዩ
ኔ፡ ፩ስሙ፡ መፍቱ
ሒምን፡ ፩ስሙ፡ ማ
ክዩስ፡ ፩ስሙ፡ ጋሮ
ን፡ ፩ስሙ፡ ወናድኤ
፩ስሙ፡ አድሂል፡ ፩
ስሙ፡ ጣምአዱዩ
ስ፡ ፩ስሙ፡ እንትዩ
ር፡ ፩ስሙ፡ ክርስቶ
ስ፡ ፩ስሙ፡ ክዱሮስ
ወአናቅጺሃ፡ ዘኗተ
እስማተ፡ ዘይዐው
ር፡ በሃውሉት፡ ወ
በአርምሞት፡ ወ
በትእጣሥት፡ ወበ
ስሆሳሰ፡ ወበፈራሃ
እግዚአብሔር፡ ያድ

ኅኑ፡ ወዘኗተ፡ ቃሰ፡
ሚያ፡ በዕዘኒሁ፡
ይሣየጥ፡ በወርቅ፡
ወበብሩር፡ ወበአ
ልባስ፡ ክቡራት፡
ወዘኗተ፡ ክርስቶስ
እመ፡ ኃጥአ፡ ይኩ
ን፡ ብእሴ፡ ግብረ፡
ስሄሐፈሁ፡ ፺ዓመ
ተ፡ ድልቅኗዖ፡ በ
ዘአጥመቀኒ፡ አባ
ዩሐኛሰ፡ በዛቲ፡ ዕ
ለት፡ ወበዛቲ፡ ሰዓ
ት፡ ይፍታሕ፡ ክርስ
ማዕፀዊሃ፡ ለአና
ቅጺ፡ ጽድቅ፡ ወኢ
ይሬእዩዋ፡ ለኗያን፡
በዛቲ፡ ቃል፡ በው
ስተ፡ ኗያኒ፡ በስ
ነ፡ ምጣባና፡ ይማ
ህሮሙ፡ እግዚአ

ብሔር፡ በዝኗቲ፡ ተ
ማኅበኒኩ፡ እንግብ
ርክ አልፋስ ኣዱ
፨ በሰመ፡ አብ፡
ወወልድ፡ ወመንፈ
ስ፡ ቅዱስ፡ ፩አምላክ፡
ኗንጣረክሙ፡ አበ
ዊነ፡ አበዊነ፡ ወአኃ
ዊነ፡ ለእለ፡ ሃእም
ኑ፡ ዘኗተ፡ ቃለ፡ ልፋ
ፈ፡ ጽድቅ፡ ዘመሀባ
ለማርያም፡ በጎል
ጎታ፡ ዕለት፡ ዘኗተ
እስማተ፡ ተዐውር
በውስተ ገጹ፡ በአ
ምሳለ ሂራ፡ ወዘሃ
ራ፡ በአምሳለ፡ ማር
ያም፡ ወዘሂ፡ እነቀ
ለዛቲ፡ መጽሐፍ፡
ኢይሬእዮ፡ ለኗያን
እስ፡ ይረክቡ፡ ሕይ

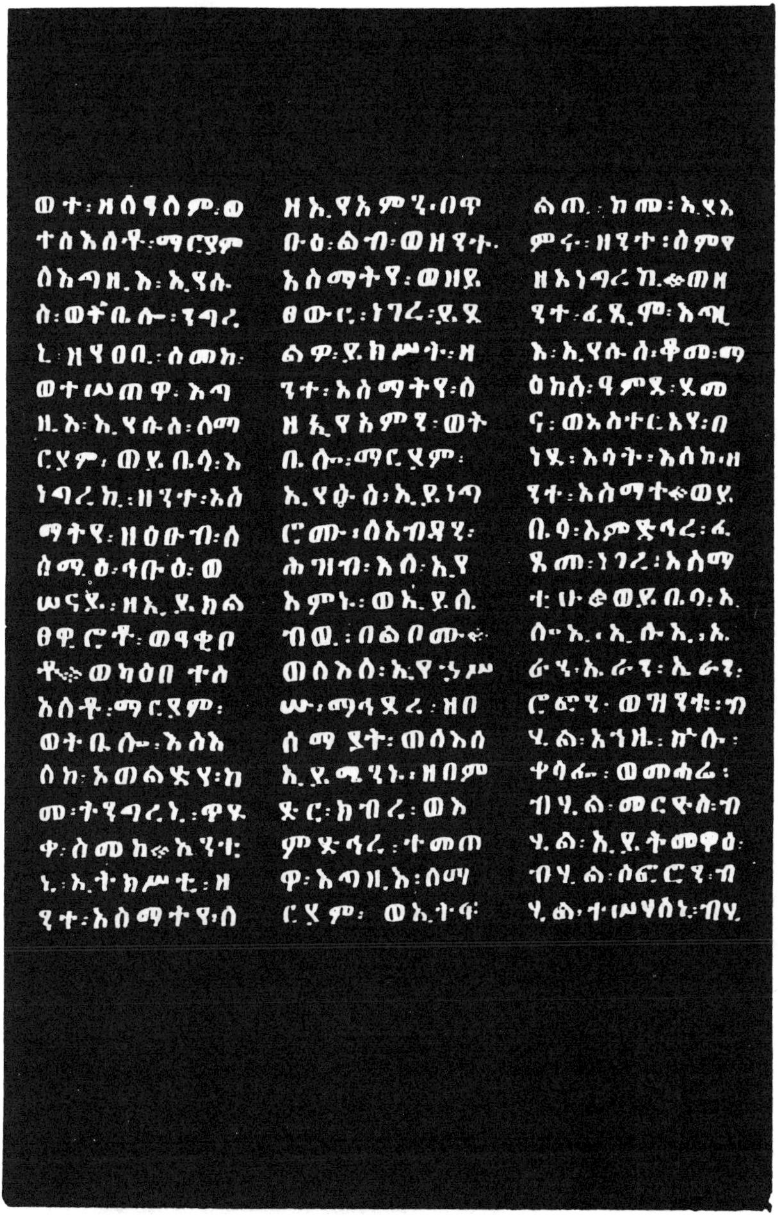

ወተ፡ዘሰናስም፡ወ
ተሰእሰቶ፡ማርያም
ሰእግዚእ፡እየሱ
ስ፡ወቶቤሎ፡ነገረ
ኒ፡ዘሃበበ፡ስመከ፡
ወተሠጠዋ፡አግ
ዚእ፡እየሱስ፡ሰማ
ርያም፡ወይቤላ፡እ
ነግረኪ፡ዘኒተ፡አስ
ማትየ፡ዘዐውገ፡ሰ
ሰማዕ፡ተባዑ፡ወ
ሠናዪ፡ዘኢ፡ይክል
ፀዋሮቶ፡መዓቂበ
ቶ፡ወካዕበ ተሰ
አለቶ፡ማርያም፡
ወትቤሎ፡እስእ
ለከ፡አመልኳየ፡ከ
መ፡ትንግረኒ፡ዋዜ
ቀ፡ስመከ፡ኢነነ፡
ኒ፡እትክሥቴ፡ዘ
ኒተ፡አስማተየ፡ሰ

ዘኢየአምኒ፡በጥ
ቡዕ፡ልብ፡ወዘየኀ
አስማትየ፡ወዘይ
ፀውር፡ነገረ፡ዩጀ
ልዎ፡ይክሥት፡ዘ
ኒተ፡አስማትየ፡ሰ
ዘኢየአምኒ፡ወት
ቤሎ፡ማርያም፡
እየሱስ፡እይነግ
ሮሙ፡ሰአብጻኒ፡
ሕዝብ፡አሰ፡እየ
አምኑ፡ወኢይሴ
ብዉ፡በልበሙ፡
ወሰአሰ፡ኢየኀሠ
ሡ፡ማኅደረ፡ዘበ
ሰማያት፡ወሰአሰ
እይሜኒኑ፡ዘበም
ጽር፡ክብረ፡ወአ
ምጽኣረ፡ተመጠ
ዋ፡እግዚእ፡ሰማ
ርያም፡ወእትፋ፡

ለጠ፡ከመ፡እየአ
ምሩ፡ዘኒተ፡ስምየ
ዘእነግረኪ፡ወመዘ
ኒተ፡ፊጸም፡እግዚ
እ፡እየሱስ፡ቆመ፡ማ
ዕከሉ፡ዓምደ፡ሂመ
ና፡ወአስተርእየ፡በ
ነጂ፡እሳት፡እስከ፡ዘ
ኒተ፡አስማተ፡ወይ
ቤባ፡እምጽኣረ፡ፈ
ጸመ፡ነገረ፡አስማ
ተ፡ሁቁ፡ወይቤባ፡እ
ሰ፡እ፡እ፡ሱእ፡እ
ራሂ፡ኤራኝ፡ኤራኝ፡
ሮፎሂ፡ወዝኒቱ፡ብ
ሂል፡አኀዜ፡ኵሉ፡
ዋሳሬ፡ወመሐሬ፡
ብሂል፡መርዩስ፡ብ
ሂል፡እይ፡ትመዋዕ፡
ብሂል፡ሰፎሮኒ፡ብ
ሂል፡ተሠሃሰኒ፡ብሂ

ል፡ ጣዎን፡ ብሂል፡
መድኃኒ፡ ኮሰ፡ ብ
ሂል፡ ስሙ፡ ለአብ፡
ሙራኤል፡ ስሙ፡ ለ
ወልድ፡ ምናቴር፡ ወ
ስሙ፡ ለመንፈስ፡ ቅ
ዱስ፡ አብጸቴር፡ በዝ
ንቱ፡ አስማት፡ ተማ
ኅበንኩ፡ እንግብርከ
ተነሥለ፡ ማርያም፡
ሚካኤል፡ ወገብር
ኤል፡ ሱራፌል፡ ወ
ኪሩቤል፡ ኡራኤል፡
ወሩፋኤል፡ ሰአሉ
ለነ፡ አስተምህሩ፡ ለ
ነ፡ ሰዳቃኤል፡ አቅና
ኤል፡ ብርሃናኤል፡
ሰአሉ፡ በእንቲአነ፡
በጸሎትክሙ፡ ከ
መ፡ ኢይድኅኒ፡ እምዝ
ንቱ፡ እሳት፡ በሰዓት፡ በ

ዝንቱ፡ አስማቲከ፡
ኒሕነ፡ አጣብርቲከ፡
አጣረ፡ ማጣ፡ ሱሩ
ቶዎኒ፡ ራሞራ፡ መ
ሲጠኒ፡ በ፲ ኡኒስሳ፡
ተማኅፀንኩ፡ በአ
ስማቲከሙ፡ ከመ፡
ታድኅኑኒ፡ ሰንብ
ርክሙ፡ ኦሌሪ፡ ስ
ነኡ፡ በኢልሄኒ፡ መ
ንበርከ፡ በማርያም፡
መዋፋሥርከ፡ ሀገር
ከ፡ ወበአርያም፡ ፋ
ኤም፡ ማኅደርከ፡
ወበማርያም፡ እምከ
ወበ፴ መኒዒ፡ ሳውዶ
ኒ፡ በ፲ወ፫ ነቢያት፡
በ፲ወ፪ ሐዋርያት፡
በ፸ወ፪ ከህናት፡ ሰ
ማዕት፡ ወበ፹ ሐራ
ሰማያ፡ በ፫ ወ፷ እር

ድሳት፡ በ፸ወ፪ ቢጽ፡
በ፫፻ወ፲ወ፰ ርቱዓ
ነ፡ ሃይማኖት፡ በ፳፬
ሊቃነ፡ መላእክት፡
ወበእልፍ፡ አዕላፋ
ት፡ ቅዱሳኒ፡ መላእ
ክት፡ ወበአስማቲ
ሆሙ፡ ከመ፡ ት
ምሐረኒ፡ ወትሣሃለ
ኒ፡ ለገብርከ፡ ተኒ
ለ፡ ማርያም፡ በጽማ
ኢል፡ ስጠ፡ ኃያል
ክ፡ በቃሌ፡ ስምከ፡
በድልቅኢ፡ ምስመ
ጥምቀትከ፡ በዘአ
ጥመቀከ፡ አበሁ
ነስ፡ በጣዑኤል፡ ሰ
ምከ፡ በዘፈትሐከ፡
አጸውተ፡ ሲኦል፡ በ
ቀተ፡ ናዊ፡ በሰተ፡ ናዊ
በቀርነሳሚ፡ ስምከ፡

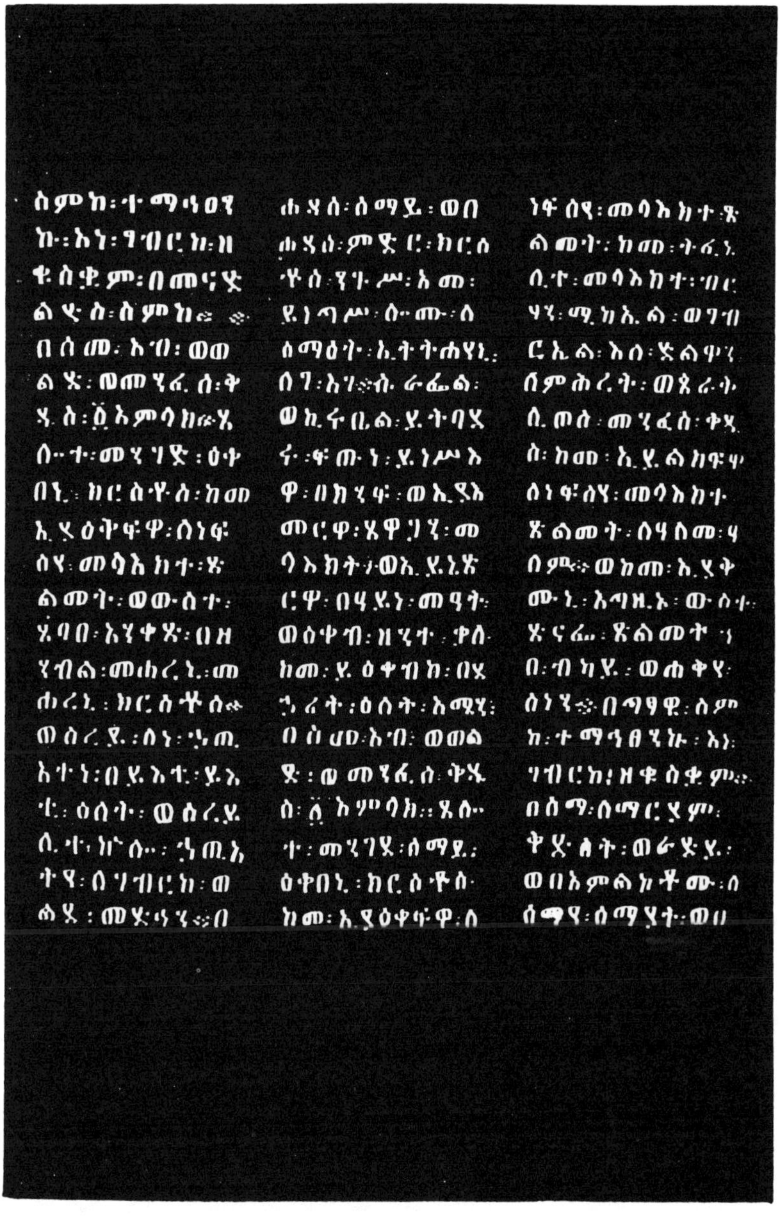

ስምከ፡ ተማኅፀንኩ
ከ፡ እነ፡ ገብርከ፡ ዘ
ቁስቋም፡ በመናፍ
ልዊ፡ ስ፡ ስምከ ። ።
በስመ፡ አብ፡ ወወ
ልድ፡ ወመንፈስ፡ ቅ
ዱስ፡ ፩ አምላክ ። ሄ
ሎ፡ ተ፡ መንግድ፡ ዕቀ
በኒ፡ ክርስቶስ፡ ከመ
ኢ.ይዕቅፍዋ፡ ለነፍ
ስየ፡ መሳእክተ፡ ጽ
ልመት፡ ወውስተ፡
ሄባበ፡ እኩይ፡ በዘ
ኮብል፡ መሐረኒ፡ መ
ሐረኒ፡ ክርስቶስ ።
ወስረዩ፡ ለነ፡ ኃጢ
አተ፡ ነ፡ በዩ ኢተ፡ ዩ ኢ
ተ፡ ዕለት፡ ወስረዩ
ሊ፡ ተ፡ ክርስ፡ ኃጢአ
ትየ፡ ለገብርከ፡ መ
ልአ፡ መዊእ ነ ። በ

ሐጻስ፡ ሰማይ፡ ወበ
ሐጻስ፡ ምድር፡ ክርስ
ቶስ፡ ንጉሥ፡ አመ፡
ይነግሥ፡ ሰሙ፡ ለ
ሰማዕት፡ ኢትትሐየኒ፡
ሰገ፡ እንግ ። ስ ራፌል፡
ወኪሩቤል፡ ይትባደ
ሩ፡ ፍ ጡነ፡ ይነሥእ
ዋ፡ በክነፋ፡ ወኢ.ያእ
መርዋ፡ ሄዋጋኒ፡ መ
ሳእክት ፡ ወኢ.ይኒጽ
ርዋ፡ በዓይኒ፡ መዓት፡
ወሰቀብ፡ ዘኒተ፡ ቅለ
ከመ፡ ይዕቀብክ፡ በጸ
ጋሪት፡ ዕለት፡ አሚን፡
በስመ፡ አብ፡ ወወል
ድ፡ ወመንፈስ፡ ቅዱ
ስ፡ ፩ አምላክ ። ሄሉ
ተ፡ መንግደ፡ ሰማይ፡
ዕቀበኒ፡ ክርስቶስ
ከመ፡ ኢ.ያዕቀፍዋ፡ ለ

ነፍስየ፡ መሳእክተ፡ ጽ
ልመት፡ ከመ፡ ትሪኒ
ሲ.ተ፡ መሳእክት፡ ብር
ሃን፡ ሚካኤል፡ ወገብ
ርኤል፡ እለ፡ ድልዋን
ለምሕረት፡ ወጸራቅ
ሲ.ጦስ፡ መንፈስ፡ ቅዱ
ስ፡ ከመ፡ ኢ.ይ.ልክፍዋ
ለነፍስየ፡ መሳእክተ
ጽልመት፡ ለዛ ለመ፡ ዛ
ለዓለ ። ወከመ፡ ኢ.ያቅ
ሙኒ፡ እግዚኡ፡ ውስተ፡
ጽናፌ፡ ጽልመተ ነ
በብክዩ፡ ወሐቅየ
ስነኒ ። በግፃዌ፡ ስም
ከ፡ ተማኅፀንኩ፡ እነ፡
ገብርከ፡ ዘቁስቋም ።
በስማ፡ ስማርያም፡
ቅድስት፡ ወራድያ፡
ወበእምልክቶሙ፡ ለ
ሰማዕ፡ ሰማዕት፡ ወበ

መንፈስ፡ ስብሐቲሁ
ለእግዚአብሔር፡ ዘ
ሐነፀ፡ ጽርሐ፡ ወአል
ቦ፡ ዘየአምር፡ ዘእንበ
ለ፡ አብ፡ ባሕቲቱ፡ ወዘ
ኢበሰ፡ ክርስቶስ፡ አመ
ሐሩ፡ መሐረኒ፡ ወሰረ
ዩ፡ ሊተ፡ ኃጢአትየ፡
ለገብርከ፡ አሌፉ፡ ስ
ኴዱ፡ በጸኃራት፡ ዕ
ሰት፡ ወበህየ፡ ይትነሥ
አ፡ ስቃነ፡ መባእከት፡
ማዕጠንት፡ ወየዓ
ርጉ፡ ጸሎተ፡ በእንተ፡
ምሕረተ፡ ሰብእ፡ ማኅ
ቶተ፡ መባእክቲሁ፡ ወ
አመ፡ ይቤልዎ፡ እግዚ
አ፡ መኑ፡ ሰብእ፡ ዘአ
የኤብሰ፡ ወዓደ፡ ዕፅ
ዘአይበደ፡ ሰ፡ ወመነ
ሰብእ፡ ዘአደ ገብር፡

ኃጢአተ፡ በጺበ፡ ም
ጽር፡ እግዚአ፡ አልቦ
ኍር፡ ዘእንበሌከ።
ወሶበሂ፡ ተናገሮሙ፡
እግዚአ፡ ኢየሱስ፡ ለ
ሰ፲ወ፪፡ ሐዋርያት፡
አመ፡ ይጽሕፍዎ፡ ለ
ዘተ፡ መጽሐፍ፡ ወ
ዴቤሎሙ፡ ለሐዋ
ርያቲሁ፡ አባሕኩክ
ሙ፡ ወዚነዊ፡ ሰሖ
ሉ፡ ዘየአምኒ፡ ብየ
በኢየሱስ፡ ክርስቶ
ስ፡ ወልደ፡ እግዚአ
ብሔር። ብፁዕ፡ ው
እቱ፡ በቃለ፡ መጽሐ
ፍት፡ ዘአጽሐፎ፡ ወ
ዘአንቀ፡ ወማየ፡ ኤሱ
ቱ፡ ሂ። ሰእመ፡ አንበ
ረ፡ ውስተ፡ ቤተ፡ ተ
ተ፡ ኢይመውት፡ ወየ

ሐዩ፡ በጸኃራት፡ ስለ
ት፡ አመ፡ ሰለተ፡ ጸደኒ
ወክሦነኒ፡ ዪቲ፡ መሐሪ፡
ወአነ፡ እምሕሮ፡ እም
ስሳተ፡ ገሃነም፡ አመ
ዴት፡ ሴለዩ፡ ኃጥአን
ወዓማስያን፡ እምየ
ጽቃን፡ እመዩ፡ መ
ዛልት፡ ወአመዩ፡ ቡ
ሰት፡ ወአሀለወት
ብዙሰ፡ ውእቱ፡ ዘሃሮ
አስመ፡ በበይ፡ ስለት፡
ዘተ፡ መጽሐፍ፡ እንተ
ትስጽጽ፡ እንዘ ኒት
ዐባሰረ፡ ሞተ፡ እም
ባሰሰ፡ ገብርከ፡ ተ
ሉ ሙረ፡ ጸጋ ወአኒትሰ
ጽ፡ ጸ፡ እምቅዋመ መ
ኒበረ፡ ስብሐቲከ፡ ሰ
ዓሰዐ፡ ዓሰም፡ አሜን
ወአሜን፡ ለይ፡ ለይ፡ ዘ፡

በስመ፡አብ፡ወወ
ልድ፡ወመንፈስ፡ቅ
ዱስ፡፩አምላክ፡ጸሎ
ት፡እግዚእትነ፡ማር
ያም፡ወሳጺተ፡አም
ላክ፡ጸሎተ፡እግዚ
እትነ፡ማርያም፡ወሳ
ጺተ፡እግዚአብሔ
ር፡ጸሎተ፡እግዚ
እትነ፡ማርያም፡ወ
ሳጺተ፡መድኃኔ፡ወ
ሳጺተ፡ሕይወት፡
ወሳጺተ፡መንፈስ፡
ወሳጺተ፡ክርስቶ
ስ፡ወሳጺተ፡ኢየሱ
ስ፡ወሳጺተ፡አማ
ኑኤል፡ወሳጺተ፡
ሃማኖ፡ወሳጺተ፡ጽ
ድቅ፡ወሳጺተ፡መፃ
ልት፡ዘኖተ፡ኵሎ
ወዘይመስሉ፡ዩሰ

ምዩ፡ወልድ፡ዋህድ፡
ቃለ፡አብ፡ቀዳማዊ፡
ዘይትነበብ፡ላዕለ
ኵሎሙ፡ሙታን፡
በጊዜ፡ግንዘት፡ወ
ሰበ፡ወረደ፡እግዚእ
ነ፡ኀበ፡አርዳኢሁ፡
አመ፡፰ሁ፡ለጥር፡
ወይቤሎሙ፡ለሰ
ም፡ለክሙ፡ጊሠመ፡
አነሥእ፡ግብረ፡ዓቢ
የ፡አማዕከሌክሙ፡
ወይቤልዎ፡መነሃ፡
ትነሥእ፡ወይቤሎ
ሙ፡ለአምዩ፡ሃረተ
ኒ፡በከርሣ፡፱አው
ራኀ፡ወአጥበወተ
ኒ፡ዋበ፡እስከ፡፫ዓ
መት፡በሐዘን፡ወበ
ምንዳቤ፡ነበረት፡ወ
ኢያእረፈት፡እሐተ፡

ዕለተ፡አነሥእ፡ወ
እሁባ፡ግብረ፡ለአቡ
የ፡ወይቤልዎ፡አር
ዳኢሁ፡እምአመ፡ሐ
ርከ፡እምነበርነ፡ኀሐ
ነ፡በሐዘን፡ወባሕቲቱ፡ን
በረት፡እምከ፡ማለ
ክሲ፡ወትናዝዘነ፡
ወትነግረነ፡በቃለ፡እ
ግዚአብሔር፡ተፈ
ሥሕነ፡ወአዕኰትና
ሁ፡ለእግዚአብሔር፡
ይእዜሰ፡ተንሥአ፡ወ
ተኃጽግነ፡እጊለ፡
ማውታ፡ንኪ፡ንወ
ይቤሎሙ፡ክርስ
ቶስ፡ዘይወሃ፡እ፡እ
ምክርሠ፡እሞጡ፡
ዩመውት፡ወአነሃ
ሞትኩ፡ወተንሣእኩ፡
ኵሙ፡ትንሥኡ፡ከማሁ፡

LEFÂFA ṢEDEḲ (B). Brit. Mus., MS. Orient. No. 551, Folio **33a**

ማርያም፡ እምሃ፡ ት
መጠውት፡ ከመ፡ ሆሱ፡
ሰብእ፡ ወዴጎረ፡ ት
ትንሣአ፡ እሙታን፡
በብዙኃ፡ ስብሐት።
ወፌኢሞ፡ ዘዓተ፡ ነ
ግረ፡ ዓርግ፡ እግዚእ
ኢየሱስ፡ ሰማያት።
ወሰቤሃ፡ ጼውዓዎ
ሙ፡ እግዚእትነ፡ ማ
ርያም፡ ስጴጥሮስ ወ
ስዮሐንስ፡ ወትቤሎ
ሙ፡ ሑሩ፡ ኀሥሡ፡
ሰዓጺ.ናተ፡ በዘትገ
ኘዙ፡ ሥጋየ፡ ወሰዛ
ኩ፡ ኀበ፡ ክሶ፡ ጸና
ጣል፡ ዘኢየሩሳሌ
ም፡ ደምጽ ኦዎኒ፡ ዶ
ኅጅራ፡ ምስሌሆ።
ወሖሩ፡ ሐዋርያት፡
ወአምጽኡ፡ ዩ.ሰኒ
ጺ.ናተ። ወአምጽኡ
ኮሎ፡ ደናግለ፡ ዘኢ

የሩሳሌም፡ ወሆሱ
ዓውዱ፡ ወአኃቲዋ፡
ማኅቶቲዊ ሆኒ፤ ወእ
ነበባ፡ መዝሙረ፡
ማኅሌት፡ ኮሰ፡ ሲሲ
ተ፡ እስከ፡ ደጸብሕ
ወነሥአት፡ እግዝ
እትነ፡ ማርያም ፄተ
ስንጺ.ናተ፡ ወአስሐ
ት፡ ውስተ፡ ምጆር፡
ወሰከበት፡ ሳዕሌሁ፤
ወሰፍሐት፡ እጺዊ
ሀ፡ ወዔስየት፡ ዘኒ
ተ፡ ጼሱተ። ወትቤ
እግዚእየ፡ እምሳከ
እስራኤል፡ እምሳ
ከ፡ ኀደሳኒ፡ እግዚእ
ብሔር፡ ዘገበርከ፡
ሰማየ፡ ወምድረ፡ ወ
ነፍሱ፡ ዘውስተ፡ ሙ
ስማዕ፡ ስእለትየ፡ ዘ
ሰአልኩ፡ ኀቤከ፡ የ
ም፡ እግዚአብሔር፡

ሰጳጕቀኒ፡ እግዚአ
ብሔር፡ ስኖኅ፡ ወሰ
ሆሱ፡ ዘምስሌሁ፡
በውስተ፡ ታቦት፡ ዘ
አድኅንከ፡ እማየ፡
አይኅ፡ ዘባረክ፡ ወአ
ብዛኅከ፡ ዘርአ፡ አብ
ርሃም፡ ስምዓኒ፡ ዮ
ም፡ ስእለትየ። ኦእ
ግዚእ፡ ዘቆምከ፡
ምስለ፡ አቡነ፡ አብ
ርሃም፡ ወምስለ፡ ዩ
ስሐቅ፡ ገብርከ፡ ዘ
አድኅንኮሙ፡ አማ
ዕከለ፡ መዴብሐኒ
ስማዕ፡ ስእለትየ፡ ዮ
ም፡ እእግዚእ፡ ዘእ
ድኅንከ፡ ሰደዕቀ፡ ገ
ገብርከ፡ እመዓተ
ሳዖ፡ ወኢሳሙ እጎ
ሁ፡ ወመራሕከ፡ ጎ
ኖተ፡ ሰሳም፡ ስማዕ፡
ስእለትየ፡ ዮም፡ ወ

ሥምረ ሐ፡ ለነፍሰ፡ ገ
ብርከ፡ እግሴ፡፡ እእ
ግዚአ፡ ዘእጁ ኂኒ
ከ፡ ለየ ሴፋ፡ ቅዱስ
ህ፡ አመፃዮ ሙ፡ ሰ
አኂዋ ሁ፡ ሐሳውኢ
ኒ ወእምበተ፡ ሞ
ቅሐ፡ ወኦሰዩ ከ በ
ትጽመ ፊሮ ፃኒ፡ ጓ
ጉ ሥ፡፡ ሰማዕ ሰአ
ለትየ፡ ሰኒተ አግዚ
አ፡ እግዚ አብሔር
ዘአጁ ኂኒከ ሙ፡ ሰ
ሐግዞብከ፡ አሰራኤ
ል፡ አምግብሮናተ
ፊሮ ሃኒ ኔጉሠ፡ ግ
ብሰ፡ ወመራከመ
በ ሰ ፡ ምሂ ሬ፡ ሮስ
ተ ዘዩ፡ ሰ ሁ፡ ሰመ
ሰማዕ፡ ጸሎትየ፡ ፉ
ም፡ ወሥምሮ ለነፍሰ
ገብርከ፡ እገሴ፡ ኦ
አግዚአ [illegible] ማዕከ

ጸሎዮ፡ ሰመሴ፡ ነቢ
ዩ፡ ወአኘፋ ዕከ፡ ዘ
ሰመ፡ ጸባዕዮ፡ ሰሐ
ግዘብከ፡ እስራኤል፡፡
ወአሮ እዩከ፡ ሽኒፋ፡
ዐበየ፡ ወኂዴሰ፡ ሰአ
ማሴቅ፡ ወምጁዲም
ወሴሳሬ፡ ወሰአማ፡
ኒጉሠ፡ ኢሃምቀ
ሰዘሰ መ፡ ነገሥተ
አሬ ማፂ፡፡ ሰማዕ፡ ጸ
ሎትየ፡ ዮምዩ፡ እእግ
ዚአ፡ ዘሰማዕከ፡ ጸሎ
ቶ፡ ለዳዊት፡ ገብርከ
ወአጁኂኒከ፡ አምአ
ዲ፡ ሳአል፡ ጸዲቂ፡ ወ
ወሀብክ፡ ኂዴሰ፡ ዓዕ
ሮ፡ ዘሰ፡ ኤሳዕቴ፡ ሰ
ማዕ፡ ጸሎትየ፡ ዮም
እእግዚአ፡ ዘሰማዕ
ከ፡ ጸሎቶ፡ ለዮ፡ ናስ
ነቢዩ፡ ወአጨ ሸ አከ
አምከርሠ ኣንበሬ፡

ሰማዕ፡ ጸሎትየ የ
ም፡ በእኒተ፡ ነፍሰ፡ ገ
ብርከ [illegible]
የ እእግዚአ፡ ዘእጁ
ኂኒከ፡ ሰሶስ፡ ፃ [illegible]
ቅከ፡ አምአዲ፡ ረበና
ት፡ አነዳኒ፡ አሰ፡ አ
ል በመ፡ ምሕረት፡
ሰማዕ፡ ጸሎትየ የ [illegible]
እእግዚአ፡ ዘሰማዕ
ከ፡ ጸሎቶ፡ ለ [illegible]
ል፡ ነቢዩ፡ ወአጁኂ
ኮ፡ እምአፈ፡ አናብስ
ት፡ ርኁባኒ፡ ወመሀገ
ከ፡ ሞገስ፡ በቅድ [illegible]
ነገሥት፡ ሰማዕ ሂ
ሰ ጸሎትየ፡ ዮም፡ አእ
ዚአ፡ ዘሰማዕከ [illegible]
ጸሎቶ፡ ለ [illegible]
ቀ አናኒ፡ ወአነ [illegible]
ዲ፡ ሚጋኤል፡ ወ [illegible]
ጁኂኒከ፡ እም [illegible]
ዮነአ [illegible] ሰት [illegible]

LEFÂFA ṢEDEḲ (B). Brit. Mus., MS. Orient. No. 551, Folio 34a

ክሂ ነያር ሰማዕ ኄ
ሱ ትሃ ሀ ም ወእኍ
ቆ ስነፍ ስ ገብር
ከ እነሴ ኦ እግዚእ
ኢየሱስ ክርስቶስ
ወልጃ እግዚአብሔ
ር ወልሁ ወፋቀር
ሃ ዘነባሬ ዪእውራ
ጌ ውስተ ከር ሠሃ
ወሮተ ሃመተ እነዘ
ተወበ እዌባዕትሃ
ወእሂት ተ እምር
ዘረከበነ መቀወ
ፉት በእነቲ እከ በ
እሄዊ ሆመ ስእየ
ሁ ጵ እከሃቻ ነሃ
ጎበሃሃ ም ተረሥ
አ ሰነፍ ስየ ወእኍ
ገና እምጸና ጌ ሐእገ
ፉንት አስዪ ኃልዕ
ዎመ ስጻ ቲ ቀ እጊ
ሰ እመሕጸወ እኍ
ገና ስነፉስየ እማሃ

እሂ ባ ወእምጽ ናፊ
ዥል መ ት አ ዎ ዐሃ ።
ኢ ሃ ነ ው ም ወ እ ም
ቫሕረ እሳት ዘእሂ
መፋ ዕ ሰ ማዕ ኩ እ
ዘ ተ ነ ነ ሮ መ ስ እ
ር ኧኢ ከ እዣ ዘ ትብ
ል ሀ ሰ ወ በ ሐ ረ እ ሳ
ት በ ው ስ ተ ፋ ን ት
ጀ ቱ ቀ ሃ ወ ኗ ዋ እ ሃ
ያ ነ ል ፉ እ ም ዓ ሌ ሃ
እ ነ ተ ው ስ ቲ ተ ፄ ከ
ሰ ማሃ ት ዘ መ ል ሰ
ል ት ሰ ማ ሃ ት ወ ዘ መ
ት ሕ ተ ም ጽ ር ተ ጀ
ሚ ሮ መ ኢ ሃ ከ ሱ
እ ጥ ፉ እ ት ሰ ው እ ቲ
እ ሳ ት አ ዓ ር እ ነ ብ
ዕ ዘ ኗ ጥ እ ሃ ጠ ፋ
ሃ ወ ረ ኢ ማ እ ገ ዝ
እ ተ ነ ሃ ፃ ዎ ም ዘ ነ ተ
ኃ ሰ ው ተ ሮ መ ተ ዓዕ
ሰ ዓ ራ ት ተ ከ ጸ ና

ሰነ ሃ ነት እነ ዘ ተ
ሐሰ ሰ በ ል ብ ወ ተ ሃ
ል እ ጠ ሞ ት ኩ ሃ
ቀ ብ ተ ነ ሠ ነ ሃ በ
ዝ ነ ት ስ ረ ጊ ር ት ከ
ጠ ዪ ም ሐ ረ ነ እ ግ
ዜ እ ብ ሔ ር አ ማ ዝ
እ ሃ ኢ የ ሱ ስ ክ ር
ስ ቶ ስ በ መ ጊ ግ ሠ
ተ ሰ ማ ግ ተ ወ ክ
ማ ሁ አ ሬ ተ ሙ ነ
ፋ ቀ ራ ነያ ር ስ ሐ ተ
ጎ ጠ ጸ ክ መ በ እ ግ
ብ ዕ እ ሰ ወ እ ነ ዚ
እ ብ ሔ ር ሰ ራ ሃ ነ
ጠ አ ተ ሃ ም ሐ ረ ክ መ
ወ ያ ጸ መ ሰ ስ እ ብ ጻ
ከ መ ወ ሀ ሃ ነ
ጠ ስ ም ው ነ ጉ ነ
ት ዘ ባ ሃ ወ እ ም ዕ ሃ
ዘ ኢ ሃ ነ ው ም ሀ ወ
ሰ እ ከ ዪ ፉ ኤ አ ግ
ኢ እ ሃ ኢ የ ሱ ስ ክ ር ሰ

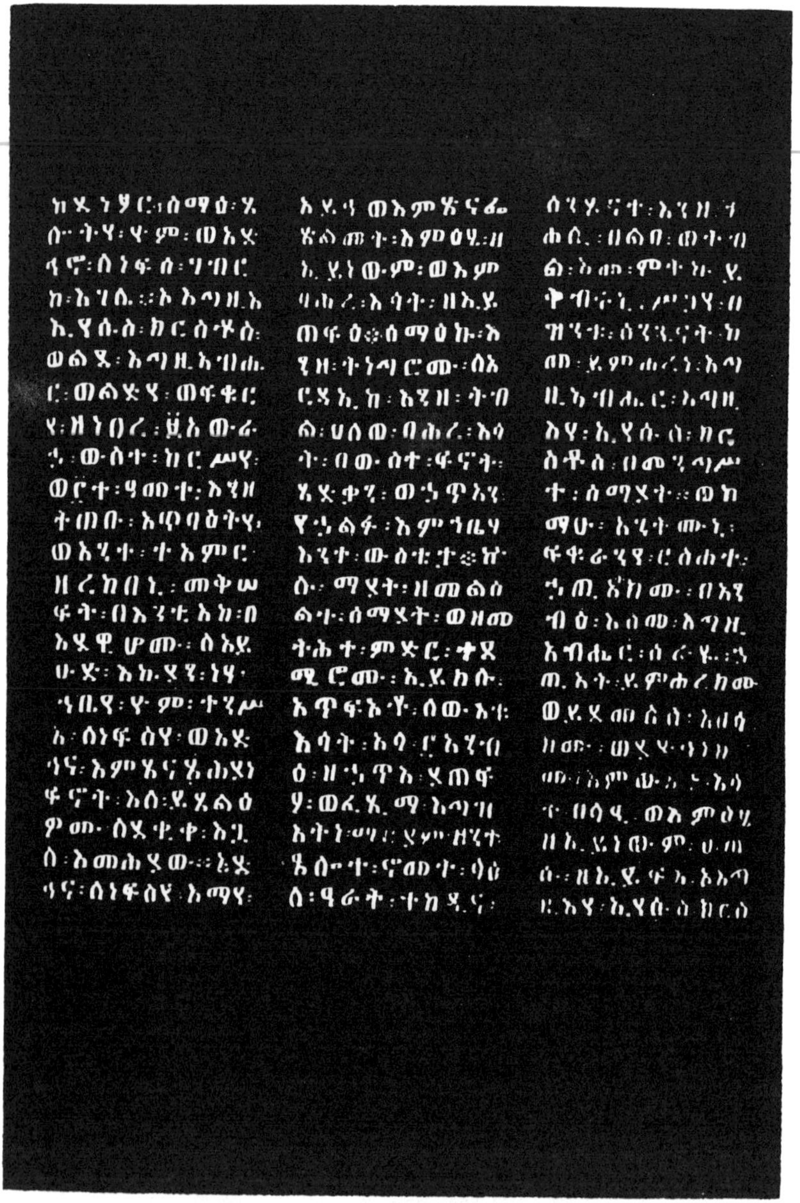

Lefâfa Ṣedeḳ (B). Brit. Mus., MS. Orient. No. 551, Folio 34b.

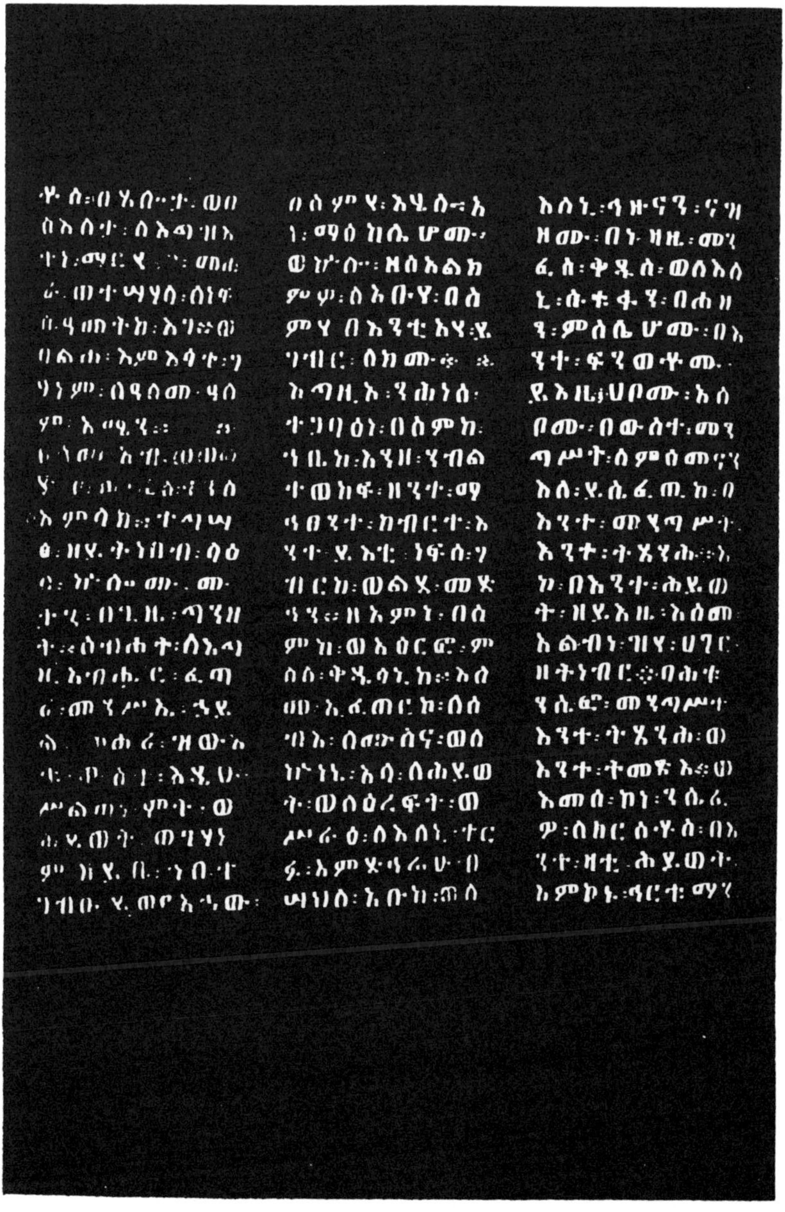